HOME

Alison Blunt and Robyn Dowling

Routledge
Taylor & Francis Group

LONDON AND NEW YORK

First published 2006
by Routledge
2 Park Square, Milton Park, Abingdon, Oxon OX14 4RN

Simultaneously published in the USA and Canada
by Routledge
270 Madison Ave, New York, NY 10016

Routledge is an imprint of the Taylor & Francis Group, an informa business
© 2006 Alison Blunt and Robyn Dowling

Typeset in Joanna and Scala Sans by
Bookcraft Ltd, Stroud, Gloucestershire
Printed and bound in Great Britain by TJ International Ltd,
Padstow, Cornwall

British Library Cataloguing in Publication Data
A catalogue record for this book is available from the British
Library

Library of Congress Cataloging in Publication Data
Home / Alison Blunt and Robyn Dowling.
    p. cm. — (Key ideas in geography)
Includes bibliographical references and index.
1. Home. 2. Households. 3. Housekeeping. 4. Dwellings.
5. Architecture, Domestic. I. Dowling, Robyn M. II. Title.
III. Series.
GT2420.B654 2006
392.3'6—dc22   2005036702

ISBN10: 0–415–33274–5 (hbk)
ISBN10: 0–415–33275–3 (pbk)
ISBN10: 0–203–40135–2 (ebk)

ISBN13: 978–0–415–33274–3 (hbk)
ISBN13: 978–0–415–33275–0 (pbk)
ISBN13: 978–0–203–40135–4 (ebk)

For Garry and Mark

CONTENTS

ILLUSTRATIONS AND TABLES

FIGURES

TABLES

Boxes

RESEARCH BOXES

ACKNOWLEDGEMENTS

This is a book that straddles our intellectual and emotional homes across two continents and brings together reading, scholarship and friendship that span fifteen years.

As a transnational endeavour, most of our collaboration occurred via email. Though email and the telephone were largely successful, working on this book was also facilitated by a couple of visits. We are grateful to the Macquarie University Visiting Scholar Scheme and the Visiting Academic programme at the Department of Geography, Queen Mary, University of London, for funding that allowed us to share our dwelling spaces occasionally and to work together in the one place.

We are very grateful to the following people who have written research boxes for the book: Akile Ahmet, Alex Barley, Sarah Cheang, Louise Crabtree, Caitlin DeSilvey, Andrew Gorman-Murray, Rachel Hughes, Caron Lipman, Katie Walsh and Helen Watkins. The boxes appear at the end of each chapter in the book, and provide vivid and engaging examples of current geographical research on home. In addition, a number of colleagues have generously allowed us to include images from their personal collections. We are very grateful to Jane M. Jacobs, Lesley Johnson, Lisa Law, Justine Lloyd, Kathy Mee, Kris Olds, and Divya Tolia-Kelly. Noel Castree, Georgina Gowans and Ann Varley also provided help and encouragement as we wrote the book.

The following gave their permission to reproduce material: Figure 2.1, Steve Brosnahan Collection of the Lower East Side Tenement Museum, New York City; Figure 2.2, *Australian Home Beautiful*, the Mitchell Library NSW and Justine Lloyd; Figure 2.3, Artangel; Figure 2.4, Michael Landy and Thomas Dane, London; Figure 2.5, Colonial Williamsburg Foundation; Figures 2.6 and 2.7,

Caitlin DeSilvey; Figures 3.1 and 4.8, Time Life Pictures / Getty Images; Figure 3.2, Ann and Thomas Damigella Collection, Archives Center, National Museum of American History, Behring Center, Smithsonian Institution; Figures 3.4, 5.1 and 5.2, Kathleen Mee; Figure 3.5, David Goldblatt; Figure 4.4, The British Library; Figures 4.5 and 4.6, Alfred de Rozario; Figure 4.7, State Library of New South Wales and Gerrit Fokkema; Figure 4.9, John Pickard; Figure 5.3, Divya Tolia-Kelly; Figure 5.4, Lisa Law; Figure 5.5, Kris Olds; Figure 5.6, Jane M. Jacobs. Box 2.9 is a revised extract from A. Blunt (2005) 'Cultural geography: cultural geographies of home', *Progress in Human Geography* 29: 505–15, and is reproduced with kind permission from Hodder Arnold. We also gratefully acknowledge permission from Blackwell Publishing to reproduce short, revised passages from A. Blunt (2005) *Domicile and Diaspora: Anglo-Indian Women and the Spatial Politics of Home* (Oxford: Blackwell). Every attempt has been made to obtain permission to reproduce copyright material. If any proper acknowledgement has not been made we would invite copyright holders to inform us of the oversight.

Robyn would like to thank her colleagues in the Department of Human Geography at Macquarie University for providing an incredibly supportive environment in which to write this book, and their commitment to the organizational change necessary to make part-time academic work possible and successful. The Australian Research Council and Macquarie University funded the research on family homes in Chapter 3 and transnational geographies of housing in Chapter 5, and in particular enabled me to gain Emma Power's exemplary research assistance. Kathy Mee and Pauline McGuirk have been very helpful sounding boards throughout the writing of this book.

On a more personal note, Garry Barrett shouldered much more than his share of our home-making practices in order for me to finish this book, and has provided unwavering support for a very long time. Eamon and Clancy handled my absences and mess in the study with their usual resilience but are nonetheless glad that Mum's takeover of the computer has ended. In a short space of time they have provided the most wonderful experiences of home as an emotional space, and for that I am especially grateful.

Alison would like to acknowledge The Leverhulme Trust for its award of a Philip Leverhulme Prize, 2004–6, which has funded her research on this book. Like Robyn, I am very glad that my departmental home in Geography at Queen Mary, University of London, is such a stimulating, supportive and collegial place to be. In particular, I would like to thank my friends and colleagues David Pinder and Jane Wills for their constant support, as well as the students on my undergraduate course, 'Geographies of Home'. Other friends too have helped me throughout the writing of this book, particularly Martin Evans, Nicky Hicks, Richard Phillips, Juliet Rowson and Elaine Sharland. Most importantly, I thank Mark Ryan and my parents, Cecily and Peter Blunt, for making home a place of love and care.

1

SETTING UP HOME: AN INTRODUCTION

What does home mean to you? Where, when and why do you feel at home? To what extent does your sense of home travel across different times, places and scales? In light of the multiple experiences of home in the modern world and the complexity of home as a theoretical concept, we surmise that we would get many and varied answers to these questions. Some may speak of the physical structure of their house or dwelling; others may refer to relationships or connections over space and time. You might have positive or negative feelings about home, or a mixture of the two. Your sense of home might be closely shaped by your memories of childhood, alongside your present experiences and your dreams for the future.

These multiple senses of home can also be seen in popular culture. Some of today's most popular television shows revolve around the minutiae of home life. Soap operas such as Neighbours and dramas such as Desperate Housewives, with their focus on suburban households, re-present to us an idea of how people relate to each other in a domestic environment. Reality programmes such as Big Brother are also about home, though filtered through the less common domestic arrangement of unrelated people living together under the constant gaze of global television audiences. And, finally, popular 'lifestyle' shows such as Ground Force, Designer Guys and The Block tell us how to make the homes in which we live more beautiful, functional and profitable. Home as a sense of belonging or attachment is also very visible in one of the key characteristics of the contemporary world: the historically unprecedented number of people migrating across countries, as, for instance, refugees and asylum seekers, or as temporary or permanent workers. Notions of home are central in these migrations. Movement may necessitate or be precipitated by a disruption to a sense of home, as people leave or in some cases flee one home for another. These international movements are also processes of establishing home, as senses of belonging and identity move over space and are created in new places.

But how do you go about understanding home? The magnitude of research on home is growing exponentially. Just in the last five years there have been special issues on home in academic journals such as *Antipode*, *Cultural Geographies* and *Signs*, as well as new journals such as *Home Cultures*, solely devoted to the subject. Books such as *Ideal Homes* (Chapman and Hockey 1999), *Burning Down the House* (George 1999), *Home Truths* (Badcock and Beer 2000, Pink 2004) and *Home Possessions* (Miller 2001), were also published during this period. Scholars from many different disciplines tackle the issue of home, each defining and understanding home in a distinct way. There is hence some confusion surrounding the term. Our aim in this book is to present an argument about home that will firstly help you navigate through this voluminous literature and secondly aid your ability to understand practices and notions of home that you will encounter in the course of your everyday life. We do not provide you with a comprehensive empirical account of home across the world and history, nor do we engage with every idea of home ever presented. Instead, this book provides a *critical geography of home*. The geographical focus of our argument certainly emanates from our training as geographers. Also important is that in geography and other disciplines such as sociology, women's studies, history and anthropology, thinking about home has been geographic, highlighting relations between place, space, scale, identity and power. In this book we develop a geographical perspective on home that uniquely identifies its two key elements. Home, as we will show throughout the book, is a *place*, a site in which we live. But, more than this, home is also an idea and an imaginary that is imbued with feelings. These may be feelings of belonging, desire and intimacy (as, for instance, in the phrase 'feeling at home'), but can also be feelings of fear, violence and alienation (Blunt and Varley 2004: 3). These feelings, ideas and imaginaries are intrinsically spatial. Home is thus a *spatial imaginary*: a set of intersecting and variable ideas and feelings, which are related to context, and which construct places, extend across spaces and scales, and connect places.

Home is hence a complex and multi-layered geographical concept. Put most simply, home is: a place/site, a set of feelings/cultural meanings, and the relations between the two. This conceptualization, which we trace theoretically in this chapter and flesh out in the rest of the book, makes explicit our key starting point: that home is much more than house or household. Whilst house and household are components of home, on their own they do not capture the complex socio-spatial relations and emotions that define home (see Box 1.1 on building and dwelling). A house is not necessarily nor automatically a home, and personal relations that constitute home extend beyond those of the household. This book, then, is about home rather than about houses, housing or households. Whilst we draw on the literature about housing and households throughout the book (and particularly in Chapter 3), we do so only where it contributes directly to a critical geography of home. There is a vast literature on housing and households that thus remains beyond the scope of this book because it does not relate so directly to understanding the socio-spatial relations and emotions bound up with home.

Box 1.1   BUILDING AND DWELLING

In his classic and challenging essay, 'Building dwelling thinking', first published in 1954, Martin Heidegger poses two questions: 'what is it to dwell?' and 'how does building belong to dwelling?' (1993: 347). Heidegger observes that building and dwelling are not the same: whilst dwellings are produced through building, not all buildings are dwellings. As he writes, 'The truck driver is at home on the highway, but he does not have his lodgings there; the working woman is at home in the spinning mill, but does not have her dwelling place there; the chief engineer is at home in the power station, but he does not dwell there' (347–8). Heidegger makes an ontological argument, whereby dwelling 'is *the basic character* of Being, in keeping with which mortals exist' (362). For Heidegger, dwelling, as mode of being in the world, is not only produced by, but also precedes building: 'Only if we are capable of dwelling, only then can we build' (362). He describes a farmhouse in the Black Forest, built two hundred years previously 'by the dwelling of peasants' (361):

> Here the self-sufficiency of the power to let earth and sky, divinities and mortals enter *in simple oneness* into things ordered the house. It placed the farm on the wind-sheltered mountain

slope, looking south, among the meadows close to the spring. It gave it the wide overhanging shingle roof whose proper slope bears up under the burden of snow, and that, reaching deep down, shields the chambers against the storms of the long winter nights. It did not forget the altar corner behind the community table; it made room in its chamber for the hallowed places of childbed and the 'tree of the dead' – for that is what they call a coffin there: the *Totenbaum* – and in this way it designed for the different generations under one roof the character of their journey through time. A craft that, itself sprung from dwelling, still uses its tools and its gear as things, built the farmhouse.

(361–2)

As Iris Marion Young explains – and as the description of a farmhouse in the Black Forest reveals – dwelling and building 'stand in a circular relation' for Heidegger, whereby 'Humans attain to dwelling only by means of building. We dwell by making the places and things that structure and house our activities. These places and things establish relations among each other, between themselves and dwellers, and between dwellers and the surrounding environment' (Young 1997: 136).

In her feminist reading of house and home, Young writes that Heidegger distinguishes between two aspects of building: cultivating and constructing. But, 'Despite his claim that these moments are equally important, Heidegger nevertheless seems to privilege building as the world-founding of an active subject' (Young, 1997: 134). For Young, 'this privileging is male-biased' (134), as women are largely excluded from the professions and trades associated with building, and often lack land title or credit to participate in building projects. In contrast, women are more often responsible for the cultivation and preservation of home through their domestic work, nurturing and care, but their importance in facilitating dwelling – and thus enabling building – is overshadowed by Heidegger's privileging of construction. As Young writes 'Those who build dwell in the world in a different way from those who occupy the structures already built,

and from those who preserve what is constructed. If building estab-
lishes a world, then it is still very much a man's world' (137; also see
Box 1.2 on Dolores Hayden's work on non-sexist house design).
Young also discusses Luce Irigaray's reading of Heidegger, which
similarly draws out the maleness of his ontological claims:

> Man can build and dwell in the world in patriarchal culture,
> [Irigaray] suggests, only on the basis of the materiality and
> nurturance of women. In the idea of 'home', man projects onto
> woman the nostalgic longing for the lost wholeness of the orig-
> inal mother. To fix and keep hold of his identity man makes a
> house, puts things in it, and confines there his woman who
> reflects his identity to him. The price she pays for supporting his
> subjectivity, however, is dereliction, having no self of her own.
>
> (Young 1997: 135; Irigaray 1992)

The cultivating, nurture and preservation of home – what, following
Young, we term home-making in this book – is a central focus of
Young's essay. She advocates a positive idea of home, which is
'attached to a particular locale as an extension and expression of bodily
routines' (161). Unlike nationalist politics that romanticize an
exclusionary 'homeland', as we discuss in Chapter 4, Young emphasizes more personal, localized experiences of home. She identifies four
'normative values of home that should be thought of as minimally
accessible to all people' (161): safety; individuation, whereby each
individual has place for the basic activities of life; privacy; and preservation. As Young concludes, 'Home is a complex ideal': whilst she agrees
with critics who regard home as 'a nostalgic longing for an impossible
security and comfort, a longing bought at the expense of women and of
those constructed as Others, strangers, not-home, in order to secure
this fantasy of a unified identity' (164), she also argues for a more positive recognition of home and its normative values. As she explains, 'the
idea of home and the practices of home-making support personal and
collective identity in a more fluid and material sense, and that recognizing this value entails also recognizing the creative value to the often
unnoticed work that many women do' (164).

Our purpose in this chapter is to establish the intellectual currents that frame geographical understandings of home. We focus on how home is defined in these intellectual traditions, what its characteristics are, and the social processes influencing it. We begin by introducing three important traditions of work – housing studies, Marxism and humanism – that have, in different ways, influenced the study of home. We then turn to two further, often intersecting, traditions – feminist and cultural geography – which are the basic building blocks of the critical geography of home that we develop throughout the book. In the second part of the chapter we present the conceptualization of home that we use in this book. We elaborate three key elements of a critical geography of home: home as material and imaginative; connections between home, power relations and identity; and home as multi-scalar.

CONFINING HOME: HOUSING STUDIES, MARXISM AND HUMANISM

Three of the key frameworks in which home is understood are housing studies, Marxism and humanism. None is solely geographical, but each has been an important influence on geographical thinking. One of the important elements of home is that it is a house or shelter; a structure in which we are housed, whether that be a tent, caravan, house, apartment, park bench, or any other assemblage of building materials on a particular site. Housing studies is a multi-disciplinary and multi-theoretical tradition that investigates the economic, cultural, sociological and political aspects of the provision of housing (browsing a couple of issues of the journal Housing Studies would be helpful here). Although housing studies is diverse and complex, it is useful to think about it as having four main strands. First, in this intellectual tradition we see a major focus on housing policy, since governments play a large role in determining what sorts of housing can be built, where it is built, its cost and forms of tenure (renting vs owning being the most common distinction). In the late 1980s, for example, there was a strong housing studies focus on changes in British housing policy that involved a shift from public to private ownership of housing (Forrest and Murie 1992). More recent interest in housing policy has turned toward investigations of the role of policy makers and ideologies in the formation of housing policy (Jacobs et al. 2003). Second, the economics of housing provision is important in housing studies. The building of new houses and the renovation of old ones generate a significant proportion of income and employment in many countries, as we shall see in Chapter 3. Whether quality housing in appropriate locations is affordable for most people remains of abiding interest. The economic characteristics of housebuilders and structure of the industry are also charted, and connections traced between the provision of housing and broad economic processes such as globalization (Forrest and Lee 2004; Olds 1995).

Third, housing studies is concerned with *house design*. In this literature, the many different forms houses take over time and across space are documented. Paul Oliver's (1987) survey of house form across the world shows the geographical and historical diversity of house design and its cultural influence, such as houses in which the location and function of rooms are related to symbolic associations of the Zodiac, or gendered associations of space (for example male spaces with light and culture, and female spaces with dark and nature). In much of this research, links between house design and social context are emphasized, with the argument being that the layout of dwellings and materials used are reflective of social and cultural norms. There is a strong tradition of feminist analysis, for example, that points out the gendered assumptions that inform much housing design, as well as the gendered practices of design professions such as architecture. Box 1.2 provides an overview of the thinking of Dolores Hayden, an architectural and cultural historian critical in understanding the past, present and future of house design. Within this design tradition houses of the future are also imagined, houses that may be more environmentally sustainable, supportive of people with disabilities, or supportive of more equal gender relations.

Box 1.2   DOLORES HAYDEN: GENDER, HOME AND NATION

Dolores Hayden's writings on home have been influential for more than two decades. From a social and architectural perspective she has made three contributions to thinking on home that we draw on throughout this book. First, Hayden provides an extended and critical examination of the links between home and nation, especially the idealization of the suburban home in American life (discussed in more detail in Chapter 3). For her, the suburban home is as much a landscape of imagination as it is a built form; embodied in this dwelling structure are social goals of individualism, privacy, idealized family life and gendered norms. For much of the twentieth century, according to Hayden, Americans' ideal life has been embodied in the house rather than the town or neighbourhood. It is also a suburban house, and one in which specific gender relations hold fast: 'the single-family house was invested with church-like symbols as a sacred space where women's work would win a reward in heaven' (Hayden 2003: 6). The important role of home in building and embodying national imaginaries is echoed in the role of home in building national *economies*. As Hayden reminds us 'the physical realization of this dream has been in the hands of developers trying to turn a profit through suburban growth' (2003: 9). From 1820 to 2000, complex negotiations between developers, residents and government produced different environments and nuanced forms of the 'dream home'.

Second, Hayden thoroughly explores the gendering of both ideals and materialities of home. Cleaning and maintaining the house requires the unpaid and unrecognized work of women (Hayden 2002). Suburban houses exacerbate the amount of work required of women because they privatize or individualize domestic labour – each household has a kitchen and laundry to clean, and separate living areas require the individual supervision of children. Suburban homes do not support communal or shared domestic labour. Just as importantly, making home depends upon myriad nurturing activities, also undertaken by women. These include cooking, care and 'home remedies' for illnesses, emotional support for family members, arranging recreational activities, establishing and maintaining ties to kin and community, and managing relationships to government (for example schools) and business (for example shops) (Hayden 2002: 82). From a feminist perspective, the home-making activities of women are problematic in a number of respects. They are largely unrecognized: 'homemakers ... have often found that the only time their work of cooking, cleaning, and nurturing compelled attention was when it was *not* done' (Hayden 2002: 84, emphasis in original). And home is often not a haven for women; it is not a space in which they can claim privacy and autonomy.

Third, and following from this, Hayden has brought non-sexist house design to the forefront of architectural thinking, outlining the long history of feminist revisionings of the suburban home (2002). Hayden draws our attention to the 1870s work of Melusina Fay

Peirce and a group of women she calls the 'material feminists', who attempted to create new kinds of homes with socialized housework and childcare. They built kitchen-less houses alongside a public kitchen and community dining club. Hayden also highlights the twentieth-century ideas of Charlotte Perkins Gilman, and in particular proposals for house design more appropriate to working women and their families (Hayden 2002: 110). Apartment houses, which included day care centres and restaurants to provide food for busy families, have been built in Stockholm and Copenhagen. Hayden herself suggests a redesign of the suburban house and neighbourhood. She proposes that if parts of individual backyards gave way to a collective village green, this would enable more shared childcare and perhaps provisioning of food through gardening (2002: 208). For Hayden, then, reform in the housing arena is a precondition of gender equality. Hayden's more recent work (Hayden with Wark 2004) turns more generally to alternative housing forms (see Chapter 3), using aerial photography to critique 'bad building patterns' and help people 'visualize positive changes' (Hayden with Wark 2004: 15).

Fourth, the *experience and meaning of home* in different forms of housing is an increasingly strong strand of housing studies. In 1988 Peter Saunders and Peter Williams asked researchers interested in the sociology and political economy of housing to broaden their concern to that of the home, which for them included 'household structures and relationships, gender relations, property rights, questions of status, privacy and autonomy, and so on' (1988: 81). There has subsequently been intense interest from housing researchers in home as an idea, in terms of experiences, feelings and meanings of home. The findings of three key articles in this field are summarized in Table 1.1. All rely on studies in the western world, and show some consistency to answers people give when asked what home means to them. Home provides shelter, and also provides a setting in which people feel secure and centred. People's sense of self is also expressed through home, though it is important to remember that these meanings of home vary across social divisions such as gender, class and race.

As you can see from this description, housing studies is theoretically and empirically wide-ranging, and we certainly draw from this tradition in this book. In particular, we draw from its focus on the importance of house as home, and use its illustrations of the social, economic, historical and cultural embeddedness of house as home. We depart, however, from housing studies' focus on housing as the sole entry point into understanding home. Home is undeniably connected to a built form such as a house, but home is not always a house. This conflation of house and home has been widely criticized. Most simply, home is a series of feelings and attachments, some of which, some of the time, and in some places, become connected to a physical structure that provides shelter. Conversely, one can live in a house and yet not feel 'at home'. A house environment may be oppressive and alienating as easily as it may be supportive and comfortable, as shown by domestic violence, 'house arrest' and home detention as alternatives to prison, experiences of poverty and poor housing conditions, and the alienation often felt by young gay men and lesbians in parental homes. Given that a house may not be a home, and that home is not only experienced in a house, in this book we explicitly explore home *within and beyond* the house. As spatialized feelings of belonging *and* alienation, desire *and* fear, the spatialities of home are broader and more complex than just housing. To understand these complex spatialities we need to move theoretically beyond housing studies to different geographical traditions.

Marxism was an important theoretical framework of human geography in the 1970s and 1980s (see Blunt and Wills 2000). At one level, Marxism-inspired studies in geography ignored the home, with their focus on production, work-places and labour processes. Home in this framework is unimportant and

Table 1.1   **Research-identified meanings of home**

Depres (1991)	Somerville (1992)	Mallett (2004)
Material structure	Shelter	House
Permanence, continuity	Hearth	Ideal
Security and control	Heart	
Refuge	Paradise/Haven	Haven
Status	Abode	
Family, friends	Privacy	
Reflection of self		Expression or symbol of self
Centre of activities	Roots	Being-in-the-world
Place to own		

irrelevant to the main focus on labour and production. Marxist frameworks do, however, include the sphere of reproduction, and here home is conceptualized as a space of social reproduction. Home is a space in which labour power is reproduced; workers fed, rested, clothed and housed. Social reproduction tends, nonetheless, to take an analytical 'back seat' in Marxist accounts (see Marston 2000). The purpose of reproduction in the space of the home is to ensure that workers are physically and emotionally able to continue working. Home also plays a role in the ideological reproduction of capitalism. The ownership of housing, in particular, encouraged commitment to, and identification with, capitalist values (Harvey 1978). The fact that workers became homeowners not only meant that they developed an economic stake in the success of capitalism but also that the financial imperatives imposed by mortgage repayments made strike action less likely. Thus the ownership of housing by the working class was a means of the incorporation of capitalist values and a reduction in revolutionary possibilities. Within Marxism, then, home was either marginalized or deemed a hindrance to progressive social change. On its own it is not a framework that can adequately underpin a geographical analysis of home.

Humanistic geographers, in contrast, place home at the centre of their analyses. With a central focus on human agency and creativity, humanistic geographers explore the ways in which places are meaningful and full of significance for people (see Cloke et al. 1991; and see Box 1.3 on the poetics of home). The theoretical lineages of humanistic geographers are multiple, and include phenomenology, humanism and existentialism. Environmental psychologists' studies of home are quite similar, as in Clare Cooper Marcus' (1995) explorations of feelings about home. Humanistic geographers focus on the meaning of home and the ways in which home is a meaningful place – how people relate to and experience their dwelling as well as how people create a sense of home in terms of comfort and belonging. Home was not necessarily a house or shelter for humanistic geographers, but was a very special place: an 'irreplaceable centre of significance' (Relph 1976: 39) and a place 'to which one withdraws and from which one ventures forth' (Tuan 1971: 189). In investigating people's experiences of the world and the places within it, humanistic geographers privilege an idea of home as grounding of identity, an essential place. For Dovey (1985), the term indicates a very special relationship between people and their environment, a relationship through which they make sense of their world. Home in this sense is much more than a house, and much more than feelings of attachment to particular places and people. Home is hearth, an anchoring point through which human beings are centred.

Box 1.3   THE POETICS OF HOME

House and home are a central focus of Gaston Bachelard's book entitled *The Poetics of Space*, which was first published in 1958. Writing from a phenomenological perspective, Bachelard's evocative reflections on intimate space have been particularly influential in humanistic writings on home and its centrality to human life. As he writes, 'The house, quite obviously, is a privileged entity for a phenomenological study of the intimate values of inside space ... [O]ur house is our corner of the world. As has often been said, it is our first universe, a real cosmos in every sense of the word' (1994: 3–4). Bachelard is concerned with the imaginative resonance of intimate spaces and their material form, as they are created, illuminated and experienced through memories, dreams and emotions. Indeed, for Bachelard, the greatest value of the intimate spaces of house and home are that 'the house shelters day-dreaming, the house protects the dreamer, the house allows one to dream in peace' (6); moreover, as he continues, 'the house is one of the greatest powers of integration for the thoughts, memories and dreams of mankind' (6). Like many humanistic geographers who have written on the home, Bachelard imagines it to be an essential place for the intimacy and creativity of human life. As we discuss elsewhere, feminist critiques have been particularly important in stressing the politics as well as the poetics of home, and rendering it a more complex place shaped by negative as well as positive emotions and experiences.

Bachelard's ideas continue to inspire humanistic writings on home, as shown by Edmunds Bunkše's personal reflections in *Geography and the Art of Life* (2004). As we discuss in Chapter 5, much of this book describes Bunkše's forced relocation from his childhood home in Latvia during the Second World War, his migration to the United States, and the pain and difficulties of returning to his original homeland. Here, however, we introduce two ways in which Bunkše develops a poetical geography of home inspired by Bachelard. First, drawing on Bachelard's ideas about 'protected intimacy in immensity' (2004: 84), Bunkše describes three homes in landscapes of alienation. In one vivid example, he writes about where he lived on his first return visit to Latvia after forty years.

Missing his wife, and feeling that he was under surveillance within as well as outside the apartment in Soviet-occupied Latvia, he writes that 'a very curious thing happened: I experienced protected intimacy, warmth, domesticity, security and homeyness in my shabby, Le Corbusier-inspired machine for living, a tiny, third-story apartment in a complex of concrete apartment blocks, brought to Riga by the Soviets' (91–2). This apartment became, to his surprise, 'one of the homiest domiciles I have ever lived in', and his 'small worktable in the living room became my essential home' (93). Describing the few hours that he spent writing at this table each evening as a refuge from the outside world, Bunkše writes that 'Home is sometimes a state of mind' (94).

Bunkše also describes the design of his current, Bachelardian home in Pennsylvania. As he explains,

> A Bachelardian home has nothing to do with what a society customarily thinks of as the ideal home; it is not a brownstone or a house in the suburbs, a domicile with so many bedrooms or bathrooms, with a parlor or a den. It has nothing to do with real-estate values or location, with being on the proverbial right side of the tracks. According to Gaston Bachelard, a home, even though its physical properties can be described to an extent, is not a physical entity but an *orientation* to the fundamental values – gathering together into 'one fundamental value' the myriad 'intimate values of inside space' – with which a home, as an intimate space in the universe, is linked to human nature.
>
> (101–2)

As we argue throughout this book, home might take the material form of a house or other shelter, but also extends far beyond a material dwelling. Bunkše's home centres around a wood-burning stove, and 'there is glass in some form on all four sides of the domicile, letting in light and the world as appropriate for each living area inside and connecting each of those areas with the outside' (108). The intimate immensity of the home reflects its location and relationships with the wider world. As Bunkše explains, his house was

> inspired by Bachelard's writings, 'as well as [by] a careful consider-
> ation of the relationship of the house (home) with the cosmos, the
> climate, the site, our cultural heritage, our life path' (105).

These geographers can be credited with putting home back on geographical agendas, for, as they rightly point out, home places and home experiences had become less visible in both spatial science and Marxism. But they can also be criticized for romanticized notions of home and for an inadequate under-standing of the relations between social structures and experiences of place. Humanistic geographers' conception of home was a static one in which home was seen to be under threat from, rather than enabled by, the modern world. From a humanistic perspective, contemporary, bureaucratic ways of organizing the world and relating to others, alongside technology and commodification, are eroding people's ability to make home, to create a place that is sacred, sepa-rate from society and full of significance (Dovey 1985). Yet home, as we outline later in this chapter, can be conceptualized as processes of establishing connections with others and creating a sense of order and belonging as *part of* rather than *separate from* society. This separation of experiences of dwelling from social structures also means that the social differentiation of these experiences goes unrecognized. In particular, the ways in which men and women, people in different societies, and the young and old, create, perceive and experience dwelling is not recognized. From a phenomenological perspective, Lynne Manzo's (2003) review of people's relationships with places includes research that moves beyond seeing home as simply a positive relationship with place, and recognizes the dynamic nature of people–place relationships. Drawing upon a broad range of social science research, Manzo illustrates the political, social and economic implications of people's relationships to place and defini-tions of home. In the next section we sketch this broader research on home in more detail, suggesting that feminist and cultural geography perspectives open up geographical understandings of home.

EXPANDING HOME: FEMINIST AND CULTURAL GEOGRAPHIES OF HOME

Feminist frameworks have been important in developing geographical thinking on home and we draw extensively upon them in this book. One of the reasons why there has been so much feminist thinking on home is because gender is crucial in lived experiences and imaginaries of home. Cutting across the diverse definitions of home used in different frameworks is a recognition that home has something to do with intimate, familial relations and the domestic sphere. Household and domestic relations are critically gendered, whether through relations of caring and domestic labour, affective relations of belonging, or establishing connections between the individual, household and society. Gendered expectations and experiences flow through all these social relations and their materialities, and gender is hence critical to understanding home (see Bowlby et al. 1997).

In addition to asserting that gender is critical in understanding home, feminists underline that home is a key site in the oppression of women. For many women, home is a space of violence, alienation and emotional turmoil. As a symbolic representation, home serves to remove women from the 'real' world of politics and business. Conservative politicians are perhaps the most well-known examples of those who define women's place in the home. As lived experience, home can be similarly oppressive. Most famously in the 1960s, the American feminist Betty Friedan described the home as an oppressive place for women, a place that confines them to the domestic sphere and does not offer self-fulfilment (Friedan 1963). As she puts it:

> As she made the beds, shopped for groceries, matched slipcover material, ate peanut butter sandwiches with her children, chauffeured Cub Scouts and Brownies, lay beside her husband at night, she was afraid to ask even of herself the silent question – 'Is this all?'
>
> (quoted in Schneir 1996: 50)

Emancipation and fulfilment for women, according to Friedan (1963), could only come from the 'public' spheres of work, politics and education; by leaving home. Friedan's solution of sending women out into the public realm was equally problematic, as we shall see in our consideration of home in terms of public and private spheres (see Tong 1989).

Feminist thought has also made important critiques of geographical thinking on home. Humanistic and Marxist frameworks have been reinterpreted from a feminist perspective. As Gillian Rose (1993) points out, humanistic geographers' characterization of home as an essential grounding of human identity is masculinist, reliant on the experiences of men rather than women. The humanist conception of home neglects women's experiences, which, as feminist research has shown, may just as likely be a place of oppression and violence as one of sanctuary and contentment (Manzo 2003). The notion of home as haven, as a sanctuary from society into which one retreats, may describe the lives of men for whom home is a refuge from work, but certainly doesn't describe the lives of women for whom home is a workplace. Marxist accounts of home have been equally criticized, especially by socialist feminists. Socialist feminists broadened Marxist research on home through their emphasis on the twin processes of patriarchy and capitalism. For socialist feminists, capitalism produces inequality in tandem with patriarchal relations and ideologies that position women as inferior to men. Thus in this framework the gendered nature of the sphere of reproduction (and hence in understanding home) must be taken into account. The domestic sphere is as much about inequality between men and women as it is about the reproduction of labour power. In addition, socialist feminists challenge the Marxist notion that the home is only a site of social reproduction, not a workplace. As we outline in more detail in Chapter 3, home can also be considered a workplace, especially for women. Domestic labour – cleaning, cooking, nurturing – occurs in the home, and though it may not be waged, it is still labour. Moreover, workplaces are not completely separate from home. Rather, home and work are connected, reliant upon one another. Most famously in geography, Mackenzie and Rose (1983) showed that historical processes of industrialization were dependent upon a domestic economy and home life (Mackenzie and Rose 1983). More recently, Susan Hanson and Geraldine Pratt (1995) have illustrated how the labour market opportunities and experiences of women are influenced by the location of their homes and the nature of their familial networks.

Articulating the links between home and work is part of a broader feminist concern to challenge and reformulate the simple categorization of home with domestic and private spheres (see Box 1.4). Many accounts of home are dependent on the separation of home from the public sphere, such as humanistic understandings of home as a retreat or refuge. Feminist research has empirically and theoretically shown that home places and imaginaries are not exclusively private, familial or feminine. The notion that public and private spheres are distinct and oppositional has been subjected to careful historical scrutiny by Leonore Davidoff and Catherine Hall (2002). In their research on the British middle classes, Davidoff and Hall suggest that the ideology of separate spheres emerged out of a particular set of circumstances from the late eighteenth century (also see Vickery, 1993; and, for a review of the wide literature on separate spheres in the United States, see Kerber, 1988). Industrialization meant that places of home and work became increasingly separate. Historical discourses of separate spheres were class-specific and clearly gendered, and

Box 1.4   DUALISTIC THINKING AND UNDERSTANDING HOME

Table 1.2   Dualistic understandings of home

Home	Work
Feminine	Masculine
Private	Public
Domestic	Civic
Emotions	Rationality
Reproduction	Production
Tradition	Modernity
Local	Global
Stasis	Change

Table 1.2 is a selective list of oppositions that have structured much thinking about home. Research on work, production and politics, for example, constructs home as outside of, and irrelevant to, these 'public' spheres. By corollary, research on home has tended to focus on women, domesticity and tradition. The ignorance of home has even occurred within architecture, a discipline with a central focus on the ways and places in which people live. Architecture's focus on public buildings such as skyscrapers, and an emphasis on form and aesthetics to the detriment of use, takes the emphasis away from residential or domestic environments (Reed 1996). These binary associations that have structured thinking on home in geography share the characteristics of binary thinking that permeates social theory (see McDowell and Sharp 1999). First, the possibility of linkages between the opposing categories is neither conceptually nor empirically considered. Second, these oppositions valorize one side and devalue the other. Thus public, work and citizenry are presented as superior to private, home and the domestic. Feminist thinking on home challenges these dualisms in a number of ways. Theoretically, dualistic thinking is criticized for inadequately understanding the porosity of categories, and the invariable connections across the opposites – between home and work, public and private, for example – are traced empirically. Examples of both these lines of critique will be found throughout our discussions.

were mobilized to help an emerging and rapidly growing middle class in Europe and North America distinguish itself in part through its domestic expectations and aspirations. Such a vision of middle-class domesticity served to elevate the privileged position of white, middle-class women over and above those women who were, and always had been, employed both inside and outside the home. As Linda McDowell writes, 'The home was constructed as the locus of love, emotion and empathy, and the burdens of nurturing and caring for others were placed on the shoulders of women, who were, however, constructed as "angels" rather than "workers"' (1999: 75–6).

Look at the painting by F. C. Witney reproduced in Figure 1.1. This was the image on the cover of a book by E. W. Godfrey entitled *The Art and Craft of Home-making*, which was published in 1913. In many ways, this image is a classic representation of separate spheres. The man is returning home from work, while his wife is standing at the door of their large house. The light and warmth of the domestic sphere seen through the door and windows suggest a safe and comfortable haven that is maintained by the wife for her husband. This image is not only gendered, but is also class-specific and infused with naturalized assumptions about heterosexual marriage and family life. The image portrays a prosperous married couple with a comfortable home and, given the date that the book was published, presumably at least one other, unseen, woman who was employed as a domestic worker. Analysing a separate-spheres ideology within historical and geographical context demonstrates the socially constructed nature of what we think of as public and private. In other words, public and private spheres are not pre-determined but rather emerge out of particular historical and geographical circumstances. We pick up this point in Chapter 3.

A second line of critique of thinking about the home in relation to separate spheres is primarily theoretical: that public and private are interdependent. Indeed, what is public has been defined through the exclusion of the private (Pateman 1989). What happens in, and definitions of, the domestic sphere are influenced by processes and characteristics of the public sphere, and vice versa. These critiques of the distinction between the public and the private have two important implications for a critical geography of home. First, they suggest that home is best understood as a site of intersecting spheres, constituted through both public and private. Since home is not simply domestic, an understanding of what home means, and how it is created and reproduced, requires as much attention to processes of commerce, imperialism and politics, for example, as to household negotiations. Second, the intersections of

Figure 1.1 Public and private spheres. Frontispiece by F. C. Witney to E. W. Gregory (1913) *The Art and Craft of Homemaking* (London: Thomas Murby and Co.).

public and private in creating homes are geographically and historically specific. Both of these insights usefully inform a geographical understanding of home, as we show in the next section.

Recent feminist thinking on home has drawn on poststructuralism and postcolonialism to build upon the more complex notion of home that emerges from a critique of the simple equation of home with a private sphere. Racialized differences have been shown to be important in experiences and conceptualizations of home. The liberal feminist conception of home as separate and overwhelmingly oppressive reflects the concerns of white, western, middle-class feminists, without interrogating the specificity of such concerns.

In response, a number of African-American feminists have been particularly influential in critiquing assumptions about the home as an oppressive place for women. As Patricia Hill Collins writes, for example,

> Because the construct of family/household emerged with the growth of the modern state and is rooted in assumptions about discrete public and private spheres, nuclear families characterised by sex-segregated gender roles are less likely to be found in African-American communities.
>
> (Collins 1991: 47)

Racialized processes of oppression through slavery and segregation produced a very different sort of home for people of colour. As bell hooks has famously written, 'homeplace' can be a site of resistance:

> We could not learn to love or respect ourselves in the culture of white supremacy, on the outside; it was there on the inside, in that 'homeplace' most often created and kept by black women, that we had the opportunity to grow and develop to nurture our spirits.
>
> (hooks 1991: 47)

In short, lived experiences of home are different for women of colour, and allow a different conception of home (and see Chapter 2 for more on the home as a place of containment and liberation). As succinctly put by bell hooks,

> At times home is nowhere. At times one knows only extreme estrangement and alienation. Then home is no longer just one place. It is locations. Home is that place which enables and promotes varied and ever-changing perspectives, a place where one discovers new ways of seeing reality, frontiers of difference.
>
> (hooks 1991: 148)

Hooks' notion of home as a fluid, mobile place that supports identities is mirrored in poststructural notions of home drawn upon by both cultural and feminist geographers. For feminists, both metaphorical and material homes support and shut down identities. Minnie Bruce Pratt's essay 'Identity: Skin Blood Heart' is often used to demonstrate the multiple spaces of home. Pratt interrogates the interplay of her white, middle-class, Christian-raised and lesbian identities at particular times and in particular places in the American

South (Pratt 1984). Her essay unsettles three homes: Alabama, where Pratt grew up and attended college; North Carolina, where she married and later came out as a lesbian; and Washington, DC, where she lives with her Jewish lover in a black inner-city neighbourhood. According to Biddy Martin and Chandra Talpade Mohanty, Pratt's essay

> is constructed on the tension between two specific modalities: being home and not being home. 'Being home' refers to the place where one lives within familiar, safe, protected boundaries; 'not being home' is a matter of realizing that home was an illusion of coherence and safety based on the exclusion of specific histories of oppression and resistance, the repression of differences even within oneself.
>
> (1986: 195–6)

Pratt's essay not only politicizes home through its focus on local histories of exploitation and struggle (Martin and Mohanty 1986: 195), but also reveals the spatial and historical variability of both metaphorical and lived homes (also see Pratt 1997).

Recent scholarship in cultural geography has explored the spatial politics of home. In her research on home and identity for Anglo-Indian women, for example, Alison Blunt studies home as 'a contested site shaped by different axes of power and over a range of scales' (2005: 4). Rather than view the home as a single, stable place where identity is grounded, this approach unsettles such ideas by pointing to the complex and politicized interplay of home and identity over space and time. As we discuss further in Chapter 4 in relation to imperial and nationalist homes, this approach also shows how the intimate and personal spaces of home are closely bound up with, rather than separate from, wider power relations.

Feminist and postcolonial-inspired studies of home within geography thus draw out a number of key elements of a critical geography of home: home as both oppressive and a site of resistance, and the many different lived experiences of home. Fundamentally, a feminist geography recognizes the fluidity of home as a concept, metaphor and lived experience.

A CRITICAL GEOGRAPHY OF HOME

In this book we use these frameworks to develop a critical geography of home. By this we first mean a *spatialized* understanding of home, one that appreciates home as a place and also as a spatial imaginary that travels across space and is connected to particular sites. Second we mean a *politicized* understanding of home, one alert to the processes of oppression and resistance embedded in ideas and processes of home. Home is a complicated term and its myriad definitions can be confusing. For purposes of clarity, we find it useful to draw out three components of a critical geography of home: home as simultaneously material and imaginative; the nexus between home, power and identity; and home as multi-scalar. These elements are not mutually exclusive but overlap. We think it useful, nonetheless, to begin with these elements individually, and bring them together in our treatment of specific lived and imagined homes in subsequent chapters.

Home as material and imaginative

Home is *both* a place/physical location *and* a set of feelings (though see Rapoport 1995 for a contrary view). As Nikos Papastergiadis (1998: 2) suggests: 'The ideal home is not just a house which offers shelter, or a repository that contains material objects. Apart from its physical protection and market value, a home is a place where personal and social meaning are grounded.' Or, as Roberta Rubenstein (2001: 1) reminds us, home is '[n]ot merely a physical structure or a geographical location but always an emotional space'. Like these authors, we insist that one of the defining features of home is that it is both material and imaginative, a site *and* a set of meanings/emotions. Home is a material dwelling and it is also an affective space, shaped by emotions and feelings of belonging. As geographers, we understand home as a *relation* between material and imaginative realms and processes. The physical location and psychological or emotional feeling are tied rather than separate and distinct. As recently put by Easthope (2004: 136): 'While homes may be located, it is not the location that is "home". Home is the fusion of a feeling "at home", sense of comfort, belonging, with a particular place'. Material and imaginative geographies of home are relational: the material form of home is dependent on what home is imagined to be, and imaginaries of home are influenced by the physical forms of dwelling. For example, if home is understood as a sanctuary from the world of work, then home will be constructed in a dwelling away from workplaces. Home is neither the dwelling nor the feeling, but the relation between the two. Indeed, as we show throughout the book, this conception of home moves us beyond the dwelling and alerts us to other sites that are called home.

Methodologically, relational geographies of home require attention to what we term home-making practices. Home does not simply exist, but is made. Home is a process of creating and understanding forms of dwelling and belonging. This process has both material and imaginative elements. Thus people create home through social and emotional relationships. Home is also materially created – new structures formed, objects used and placed. According to Inge Maria Daniels (2001: 205) 'the material culture of the home is expressive of the changing social relationships of its inhabitants [and illustrates] the complexities, conflicts and compromises involved in creating a home'. In charting home as a material and social process, we draw upon Daniel Miller's work (outlined in Box 1.5) and also the book *Home Rules* (1994) by Denis Wood and Robert Beck. Like Miller and Daniels, Wood and Beck conceive that things, and the material structures of dwellings, 'embody the values and meanings that made, selected, arranged, and preserved them' (xvi). Values and meanings are expressed through rules that prescribe what can and cannot be done within the home. For Wood and Beck, 'without these rules the home is not a home, it is a house, it is a sculpture of wood and nails, of plumbing and wiring, of wallpaper and carpets' (1). Home is lived; what home means and how it is materially manifest are continually created and re-created through everyday practices.

Box 1.5   DANIEL MILLER: MATERIAL GEOGRAPHIES OF HOME

Our understanding of material geographies of home is inspired by the work of anthropologist Daniel Miller. Spanning a couple of decades, Miller's research has focused on the cultural practices that occur within, and create, house as home. Most significant for the argument of this book is his re-development of the material culture perspective. Miller advocates returning our gaze to objects or things. Things matter, in the simplest sense, Miller says, because our social worlds are constituted through materiality (1998: 3). Objects such as banners or a soft drink bottle, for example, embody and communicate cultural meaning. But things also matter more subtly and mundanely than simply symbolizing identities or processes. Objects and things are 'employed to become the fabric of cultural worlds' (6) and 'through dwelling upon the more mundane sensual and material

qualities of the object, we are able to unpick the more subtle connections with cultural lives and values that are objectified through these
forms' (9). This is not a book about the material culture of the home,
so we do not comprehensively use Miller's suggestions on how to
study material domains. Nonetheless, we do pay attention to what
things in the home (indeed the dwelling itself) are doing in social
terms.

A second aspect of Miller's research that informs our critical
geography of home is his understanding of the relationship between
people and the material and socio-political structure of the dwelling.
Miller has formulated concepts such as appropriation and accommodation to describe and explain the processes by which dwellings
are personalized and, more profoundly, the ways in which home is a
process (see Miller 2002 and our discussion of *Home Rules* above).
In this way our attention is drawn not only to the objects we place in
our dwellings, but also to the ways in which we use them to create
home and its social differentiations.

Home, identity and power

Home as a place and an imaginary constitutes identities – people's senses of
themselves are related to and produced through lived and imaginative experiences of home. These identities and homes are, in turn, produced and articulated through relations of power. We find Doreen Massey's writing very
helpful in understanding home, identity and power. Massey (see Box 1.6)

Box 1.6   DOREEN MASSEY: A PLACE CALLED HOME

Doreen Massey has been a key figure in conceptualizing home as a
place and in terms of geographical scale. Home has been an implicit
focus of much of Massey's work (for instance on the gendered
constructions of home and work for employees in the high technology sector, see Massey 1995a) but became explicitly considered
in the early 1990s. In 'A Place Called Home' Massey (1992) outlines
the common argument that places and our sense of home are fast

disappearing in an era dominated by the global corporation, global media flows and a seeming lack of attachment to place. Massey critiques this argument because of its interpretation of places as 'singular, fixed and bounded entities defined through opposition' (12). Instead Massey reminds us that

- 'a place is formed out of the particular set of social relations which interact at a particular location' (12);
- some of these 'social interrelations will be wider than and go beyond the area being referred to in any particular context as that place'. In relation to home 'a large component of the identity of that place called home derived precisely from the fact that it had always in one way or another been open; constructed out of movement, communication, social relations which always stretched beyond it' (13);
- the identity of place is in part constructed out of *positive* (rather than negative) inter-relations with elsewhere;
- places are shaped by power geometry, whereby 'different social groups, and different individuals, are placed in very distinct ways in relation to these flows and interconnections' (Massey 1991: 25); and
- 'the crisscrossing of social relations, of broad historical shifts and the continually altering spatialities of the daily lives of individuals, make up something of what a place means, how it is constructed as a place' (Massey 2001: 462).

Massey's rich conception of place – as intersecting social relations, as open and porous – guides the conceptualization of home we present in this book. We also draw upon Massey's more recent emphases on the embodied and affective nature of home. In recalling her family's experiences of their home in Wythenshawe (Massey 2001) and through the prevalence of nostalgic constructions of home in her work (see Massey 2005), she reminds us that home is also an emotional place.

writes that homes, and other places, have a 'power geometry' whereby people are differently positioned in relation to, and differentially experience, a place called home. For example, home for women, as feminists have shown, is more likely to be associated with feelings of isolation than for men. The 'power geometry' of lived and imagined homes is such that a dominant ideology of home valorizes some social relations and marginalizes others; defines some places as home and others as not, some identities homely and others not, and some experiences at home alienating, others fulfilling. The dominant meanings of home become most apparent in Chapter 3. They include family, patriarchal gender relations, stability, security and owned dwellings. Our concern in this book is to interrogate rather than accept such normative notions of home, explicating how normative imaginaries of homes are mobilized in social processes. We also explore the contestation and reworking of home, and the transgression of dominant identities, by different social groups that unsettle such normative notions of home.

Homely and unhomely homes are a central focus of this book. We use the terms first to signal processes whereby certain dwellings and experiences are cast as homely, as shown by the design and experiences of suburban homes by and for families that are assumed to be nuclear and heterosexual. In contrast, other dwellings and experiences – such as living in a refugee camp or a homeless hostel – may appear 'unhomely' since they do not correspond to normative notions of home. Throughout the book, we carefully dismantle the dichotomy between homely and unhomely homes. We do so by exploring the ways in which apparently homely places might be unhomely and places that are assumed to be unhomely are rendered homely. To do so we draw on Sigmund Freud's influential work on the uncanny. For Freud, the term 'heimlich' refers to familiarity and homeliness, whereas 'unheimlich' refers to unfamiliarity and unhomeliness. But, rather than exist in opposition to each other, 'one seems always to inhabit the other' (Gelder and Jacobs 1998: 23; also see Vidler, 1994). As Ken Gelder and Jane Jacobs explain, 'An "uncanny" experience may occur when one's home is rendered, somehow and in some sense, unfamiliar; one has the experience, in other words, of being in place and "out of place" simultaneously' (23). Just as the homely can be rendered unhomely, unhomely places may become homely, as we show in our discussion of sites like suburban homes and homeless shelters.

Home as multi-scalar and open

The final component of a critical geography of home relates explicitly to the spatialities of home. Here we draw from Doreen Massey's conception of place outlined in Box 1.6 and Sallie Marston's expositions of geographical thinking on scale introduced in Box 1.7. Home as a place is a porous, open, intersection of social relations and emotions. As feminists have pointed out, home is neither public nor private but both. Home is not separated from public, political worlds but is constituted through them: the domestic is created through the extra-domestic and vice versa. It is certainly the case that familial relations and identities are important in imagined and lived homes, as you will see throughout this book. But it is also the case that other social relations also intersect in a place called home. For example, Lesley Johnson (1996) compellingly shows that modernity is also constituted through home. Experiences within the home – such as caring for children or older relatives, and relations with neighbours – are critical in the creation of citizens and modern subjects. Or, as Sallie Marston (2000: 234) reminds us, home-based consumption of goods such as furnishings, clothing and food also involves the construction of class, national and diasporic identities (see in particular Chapter 4 on home, nation and empire, and Chapter 5 on transna-tional homes). In sum, given that multiple social processes intersect in and consti-tute home, then it also follows that through home, multiple identities – of gender, race, class, age and sexuality – are reproduced and contested.

A second spatiality of home is that home is multi-scalar. Much of the litera-ture on home that we have reviewed in this chapter focuses only on the dwelling and/or household. In contrast, for us senses of belonging and alien-ation are constructed across diverse scales ranging from the body and the household to the city, nation and globe. Sallie Marston's (2000) clear expla-nation of the social construction of scale provides helpful theoretical and empirical signposts here (see Box 1.7). We take this recognition of scales as

---

Box 1.7   SALLIE MARSTON: HOME AS A SITE OF SOCIAL REPRODUCTION AND A KEY GEOGRAPHICAL SCALE

Sallie Marston implores geographers to consider the ways in which home and household are key to an understanding of *geographical scale*. For Marston, like other theorists of scale, geographical scale is socially produced, or made. Scale is 'not a preordained hierarchical nomenclature for ordering the world', but 'made by and through social processes' (2004: 172). Scales – like home, city, nation and empire – are, according to Marston, neither pre-determined nor mutually exclusive. Instead, a scale is constructed through *inter-rela-tions* with other scales. Geographical thinking on scale in the 1990s, according to Marston, focused on the role of economic and produc-tion processes in constructing scales. Marston argues that such accounts ignore social reproduction and in so doing are unable to grasp the complexity of the social construction of scale (Marston 2000). For Marston, the social, physical, cultural and emotional infrastructure provided in and by households not only is central to the maintenance of capitalism, but also connects with, and constructs, the scales of home, nation and city.

Marston uses the example of the domestic sphere in the United States in the nineteenth century to demonstrate the social construc-tion of home/household as the scale of social reproduction, as well as the porosity of home. During this period, notions of production and efficiency came to define the home:

> Increasingly women came to regard the home as a sort of small-scale manufacturing site with directly delivered utilities and new technologies and products reducing the need for live-in servants. These innovations had the impact of transforming women's roles within the household, as they came to take more of a direct responsibility for housekeeping and, in the process, saw themselves as professional domestic managers.
>
> (Marston 2000: 236)

On the other hand, and simultaneously, home-based practices influ-enced scales beyond the home, and especially the urban and national scales. In treatises on domestic management, and in domestic practices of the day, home-making was characterized as a public function and therefore constitutive of the public sphere. The provision of a hygienic home in a managerialist manner was seen to create better communities and an essential foundation of modern cities. Moreover, home became a kind of 'new democracy' in which women defined not only their political roles as citizens but also a new version of citizenship. Home-making practices served to create women as *productive* citizens of the state, with associated rights and responsibilities. In sum, the late-nineteenth-century home became a key site of national political engagement for middle-class women (see also Marston 2004 on this point).

multiple and overlapping in two directions in our critical geography of home. First, home places do not have to be a house or a dwelling, although they often are. Imaginaries of home and home-making processes may also construct home as occurring in and through other scales. Home places can be a suburb, neighbourhood, nation, or indeed the world. Second, imaginaries of home occur in and construct other scales as well. Feelings of belonging and relations with others could be connected to a neighbourhood, a nation, stretched across transnational space, or located on a park bench. Conversely, the scale of the domestic is implicated in processes constructing and occurring through other scales. As our discussion of Marston's work illustrates, the characteristics of the nation, and the nature of national politics, can be influenced by processes occurring at the scales of home and household. The structure of the book reflects the importance of home over different scales as we move from a household scale, through the intersections of home, nation and empire, to consider home on a transnational scale.

STRUCTURE OF THE BOOK

This book thus charts a critical geography of home. We begin with a chapter on the wide range of sources and methods that are important for studying home, particularly life stories, literature, and a variety of visual and material cultures. The following three chapters each focus on a specific scale of home: house-as-home; home, nation and empire; and transnational homes. Read together, the chapters illustrate the multi-scalar notion of home. Individually, they are designed to deploy each of the components of a critical geography of home. Thus in Chapter 3 – house-as-home – we illustrate the development and mobilization of a normative imaginary of home, its manifestation in the 'ideal' suburban home and its reproduction of patriarchal and heterosexist gender relations. This chapter also foregrounds resistances to, and transformations of, normative homes, within and beyond the suburban home. Chapter 4 turns to imperial, nationalist and indigenous ideas of home and home-making practices. We consider material cultures of imperial domesticity and the domestication of imperial subjects as well as the mobilization of the home as a site of nationalist resistance and its importance in the contemporary politics of homeland security. We also discuss the importance of home – and the possibilities of thinking about the nation-as-home – in relation to indigenous politics. Finally, in Chapter 5 we turn to the material and imaginative geographies of transnational homes, focusing on the ways in which the politics, lived experiences and conceptualizations of home for transnational migrants are mobile and multiple rather than fixed and static.

Within each of these chapters we draw out a number of key themes that correspond to our three components of a critical geography of home. Each chapter addresses the *materiality* of home, especially through domestic architecture. But because material and imaginative elements of home are relational, our discussion of domestic architecture emphasizes its connections to feelings of belonging, alienation and its constitutive social relations. An interest in the diverse *politics* and power geometries of home also runs throughout the chapters. This is not only in terms of formal politics (as, for example, the role of the state in establishing norms of ideal homes (Chapter 3), protecting the homeland (Chapter 4), and housing refugees and asylum seekers (Chapter 5)) but also in terms of the political contestation and renegotiation of homes and identities over domestic, national and transnational scales.

Throughout the book, we draw on particular examples to illustrate and develop our argument. Choosing these examples has perhaps been the most difficult task we have faced. In explaining our choice of examples we would like to remind you what this book is not: it is *not* about housing and households, nor is it an exhaustive overview of scholarship on home. Instead, our aim is to develop a critical geography of home. Thus our first criterion in selecting examples has been the extent to which they illustrate the concept of home that we have outlined in this chapter. A second but equally important principle is to draw on examples from as wide a range of places and times as we can. Thus in Chapter 3, we focus primarily on house-as-home in relation to Australia, Europe and North America as we explore the ways in which western domesticity is bound up with wider ideas about western modernity. In Chapter 4 we begin by discussing imperial domesticity on the American frontier and in British India, before exploring the importance of the home in anti-imperial nationalist politics in India and in the contemporary politics of homeland security in the United States. Chapter 4 ends by turning to contemporary indigenous politics about home and belonging in Canada and Australia. Chapter 5 draws on a wide range of examples that include the global South Asian, Irish and Chinese diasporas, refugee camps in Lebanon, the repatriation of refugees to Ethiopia, and what it means to live in a global city or suburb in multi-cultural countries such as Australia. Our use of examples in Chapter 6 is somewhat different. We end the book by showing how a critical geography of home – the argument that we develop throughout the book – can be used in relation to two contemporary, and very different, case studies:

the dismantling and reconstitution of home and shelter associated with disasters and new visions of home materialized through collaborative housing. Whilst this chapter has *set up home* by introducing a range of different ideas and approaches, the last chapter *leaves home* by providing points of departure for your own research.

Throughout the book, we also focus on some key ideas or examples in boxes. We don't want to convey the impression that this material is in any way tangential to the argument of the book, or that it can be easily contained and bounded. Rather, we want to do the opposite, and have included these boxes to focus in on certain key topics. Also scattered throughout the book are 'research boxes', which describe a wide range of recent and current research by graduate students on the theme of home. Some students have completed their PhDs already, whilst others are in their first year. Some of the boxes therefore discuss completed projects, whilst others are more prospective, considering research questions and research design. We include these boxes to showcase current geographical research on home and to highlight its rich diversity of questions, contexts and approaches.

2

REPRESENTING HOME

From architectural plans to particular objects within the home, and from interviews and ethnographic research to textual, visual and statistical analysis, the sources and methods used to study the home are diverse and wide-ranging. Whilst Chapter 1 introduced the reasons why geographers and other researchers have become increasingly interested in studying home, this chapter focuses on how research on past and present homes has been conducted. The chapter has two main aims. First, it introduces methods and sources that have been important in work that analyses the spatiality of home. Second, the chapter aims to encourage you to think about the different ways in which you might want to study home. We explore the sources and methods used to study the material and imaginative geographies of home, and the ways in which the home is politically, socially and culturally constituted, but lived and experienced in personal ways.

The chapter has four parts. It begins by exploring life stories of home in life writings and in interviews and ethnographic research. We then turn to consider various traditions of writing home, focusing on novels, magazines and household guides. We focus next on visual and material cultures of home, exploring the home as both subject and site in art, design and other media, and the use, display and meanings of objects within the home. Finally, we consider the use of quantitative research in studying home through the analysis of data sets and household surveys. Throughout the chapter, we draw out key concerns that have inspired and shaped different methodological approaches to studying home. These concerns include a recognition of the home as an intensely political site; the desire to make visible the spaces and social relations of home and domestic life that often remain less visible than public spaces and social relations; the ways in which the home and domestic life are often associated with women and an idealized notion of the family; and the importance of unsettling the home as a fixed and stable location.

LIFE STORIES OF HOME

The term 'life stories' is broad, encompassing the study of people in their own words, both in text (including diaries, memoirs, letters and autobiographies) and in person (through, for example, interviews, focus group discussions, reminiscence work and ethnographic research) (Blunt 2003a). Across this wide range of biographical forms, life stories provide a rich source for studying personal memories and lived experiences of home. But such memories and experiences, and the life stories that describe them, are never solely personal. They should rather be interpreted in relation to the wider political, social and cultural contexts within which they are situated and constituted. In this section we consider a range of historical and contemporary life stories that explore the complex connections between home, memory, identity and everyday life.

Home and identity in life writings

Personal narratives about home provide vivid accounts of everyday, domestic life and about what it means to feel at home or not at home, both in the past and the present. In *Dwelling in the Archive*, Antoinette Burton poses important questions about the political importance of the home as a site of history and memory:

> What counts as an archive? Can private memories of home serve as evidence of political history? What do we make of the histories that domestic interiors, once concrete and now perhaps crumbling or even disappeared, have the capacity to yield? And, given women's vexed relationship to the kinds of history that archives typically house, what does it mean to say that home can and should be seen not simply as a dwelling-

> place for women's memory but as one of the foundations of history –
> history conceived of, that is, as a narrative, a practice, and a site of desire?
>
> (Burton 2003: 4)

Memoirs, diaries, letters and other personal writings are a particularly rich source for studying 'private memories of home'. Such narratives are often seen as private and domestic, both in their content and in the material conditions of their production, and often document the lives, experiences and feelings of women that usually remain hidden in more public historical narratives. As Harriet Blodgett explains, letters and diaries have been the most common form of women's writing for centuries, 'expressing a resilient creative impulse that through serial writing could find outlet in a sanctioned form' (Blodgett 1991: 1). Whilst many diaries by women document the domestic details of everyday life, others record domestic disruption at times of war (see, for example, Anne Frank's diary (Frank 1967), which documents daily life for this Jewish girl and her family, who lived in hiding in occupied Holland during the Second World War).

Contemporary as well as historical life writings have been particularly important in exploring geographies of home in relation to gender, 'race', class and sexuality. Such life writings tell spatial as well as temporal stories, often charting material as well as metaphorical understandings of home and raising important questions about the politics of home. Minnie Bruce Pratt's essay that we referred to in Chapter 1 vividly illustrates the multiple spaces of home and memory. Other autobiographical writings have also explored racialized understandings of home and identity. As Becky Thompson and Sangeeta Tyagi, the editors of a book entitled Names we Call Home, explain, 'For all of the contributors, writing about racial identity required revisiting our homes of origin – our families, neighborhoods, and childhood communities' (Thompson and Tyagi 1996: xv). But, rather than situate such 'homes of origin' solely in the past, the contributors reflect on their resonance over time and space in relation to racial politics and their racialized identities. Reflecting on their deeply personal memories of childhood homes, they explore not only the ways in which the politics of both home and identity extend far beyond an individual site of origin, but also the ways in which such origins are themselves unsettled. In different ways, these personal writings 'carve out the creative spaces we need to maintain antiracist stances in our work and lives' (ix), revealing multiple identities that resonate far beyond an individual subject, and exploring creative spaces of home that extend far beyond a singular, static and bounded location. Similar themes are also important in a wide range of life writings about both forced and voluntary displacement through migration, exile and dispossession (see our discussion of Edmunds Bunkše's memoir in Chapter 5; also see Hoffman 1989; Morgan 1990; Alibhai-Brown 1995; Lim 1997).

Oral history and other interviews

Life writings about home reveal personal and often hidden stories that resonate over shared and more collective terrains. In similar ways, life stories in person are often collected and analysed to explore personal memories and experiences of people whose lives might remain marginalized or invisible and subjects – such as the home and domestic life – that might be absent from other sources. As Perks and Thomson (1998: ix) explain, '[oral history] interviews have documented particular aspects of historical experience which tend to be missing from other sources, such as personal relations, domestic work or family life, and they have resonated with the subjective or personal meanings of lived experience' (also see Gluck and Patai 1991; Thompson 2000; Blunt 2003a). In oral history interviews, other in-depth interviews, and in a wide range of ethnographic research, studies of home span the memories and meanings of home, domestic work and family relationships.

Oral histories of home exist as both primary and secondary sources, whereby researchers either conduct interviews themselves and/or consult collections of previously recorded interviews in libraries and on websites (for more on how to conduct oral history interviews, see the webpages of the British-based Oral History Society – www.oralhistory.org.uk – and its journal *Oral History*). Oral history interviews are particularly revealing in learning about how homes and domestic life have changed over time focusing, for example, on domestic work, domestic technologies, household structure and family relationships (including Parr 1999), and about a wider sense of feeling at home or not at home through, for example, narratives of migration and displacement (including Webster 1998; Walter 2001; Blunt 2005).

Local history groups often collect oral histories to document change over time and personal experiences of living in particular places, homes and communities. The work on 'Hidden Histories' by Eastside Community Heritage, for example, includes oral history projects documenting the lives of 'ordinary' people in East London. Projects include 'Green Street Lives', which is about one street in Newham since the 1960s; 'Stories from Silvertown', which is run with Silvertown Residents Association; and 'The Teviot Estate', which is about experiences and memories of living and making home in a postwar housing estate in Poplar, London (www.hidden-histories.org/esch_pages. Also see McGrath 2002).

A number of historical geographers have analysed oral history narratives alongside other sources (also see Jacobs 2001, on the importance of historical research in housing studies). Mark Llewellyn (2004a), for example, explores lived experiences alongside the 'envisioned spaces' of Kensal House in London, which was the first housing estate inspired by Modern architecture to be built in Britain. In his 'thick description of home' (229), Llewellyn quotes from oral history interviews with the remaining three original residents who still live in Kensal House, alongside an analysis of plans and designs, a Mass-Observation survey conducted in 1939, and two documentary films (the Mass-Observation surveys on housing in the late 1930s and 1940s contributed to 'An Enquiry into People's Homes' published in 1943 (Mass-Observation, 1943). Visit www.massobs.org.uk for more information on the Mass-Observation archives, which are held at the University of Sussex). As Llewellyn writes, 'we … need to chart the dialogue between architects, planners, and inhabitants on the domestic modernities that were produced, reproduced and consumed' in order to recognize that 'homes are no longer just dwellings but are untold stories of lives being lived' (246; see Box 2.1 on house biographies and Box 2.2 on housing histories).

Other life stories about home are available on the internet, both in written and oral form. For example, the website of the Irish Centre for Migration Studies at University College Cork includes recordings of oral history interviews from three projects, each of which relates to ideas about home, belonging and displacement. 'Breaking the silence: staying at home in an emigrant society' documents the often neglected impact of large scale migration on those people who remained at home; 'The Scattering: Irish migrants and their descendants in the wider world' explores experiences of migration and life in the Irish diaspora; and 'Immigrant lives: eleven stories of immigrants in contemporary Ireland' records more recent experiences of migration and resettlement (http://migration.ucc.ie/oralarchive). Oral history narratives are also represented in other forms too, including museum exhibitions, reminiscence work and a wide range of performances. The Maine Feminist Oral History Project in the United States, for example, was organized in 1992 'to collect and preserve the histories of feminists of the 1960s and 1970s and of the organizations they founded to effect social change'. This included collecting the oral histories of 29 founders and other feminists involved in the

Box 2.1   HOUSE BIOGRAPHIES

The term 'house biographies' refers to telling the story of a house-as-home (see Chapter 3) through the lives of its past and present inhabitants. The home is thus interpreted as a site of history and memory, and is brought to life through the histories, memories, imaginations and possessions of its residents. In fiction, for example, Antoinette Burton (2003) explores Attia Hosain's novel *Sunlight on a Broken Column*, which represents the trauma of the partition of India through the 'biography of a house'. As a contemporary example of a non-fictional house biography, Julie Myerson's book *Home: the Story of Everyone who ever Lived in our House* (2004) traces the lives of eighteen generations of residents since her house in Clapham, south London, was built in 1872 (also visit www.homethestory.com). Myerson's book details her research in public record offices, family record centres and probate search rooms, and her correspondence and telephone calls to people who used to live in her house. The house itself is not only a repository of memories of past lives and experiences but is also understood through such memories and experiences. Past residents of 34, Lilleshall Road were diverse, as a review summarizes:

> There is Leon, whose father moved to number 34 from down the road after his marriage broke up, and who was allowed Tottenham Hotspur wallpaper in his bedroom; Reggie, the spivvy car-mechanic who bought the house when he returned from the war and turned it into flats, evicting the tenants whose homes he had promised were secure; Lucy, the maiden aunt who was a buyer for the department store, Derry and Tom's; Edward Maslin, the original owner, who was a devoted servant to Queen Victoria. There are children who die at birth, children who are unwanted, war-time lovers, bigamists.
>
> (Wilson 2004)

Myerson's home also tells a story of migration. Alvin Reynolds migrated from the West Indies and lived in the house from 1959 to

1979, and Doreen Webley moved in as a sixteen-year-old in 1978 to join her mother who had left Jamaica when Doreen was two.

The Lower East Side Tenement Museum in New York City also tells stories about home and migration over time. The Museum is based in a tenement house at 97 Orchard Street, where more than 7,000 people from over twenty countries lived from 1863 to 1935 (Lower East Side Tenement Museum 2004). In the early years, most residents were German Protestants and Catholics, but, by the 1890s, the increasingly crowded tenement house was home to Russian, Austrian, Polish and Romanian Jews. Residents were evicted in 1935 because of new housing regulations, and the apartments remained empty until the Museum opened in 1988. Visitors to the Museum go on tours of a number of reconstructed apartments that focus on particular families who lived in the building at different times (see Figure 2.1, and visit www.tenement.org for a virtual tour of 97 Orchard Street). The history of the tenement house is thus told through the life stories of inhabitants such as the Gumpertz family from Prussia who moved to Orchard Street in 1870, the Confino family of Sephardic Jews who moved from Kastoria in the Ottoman Empire to New York in 1913, and the Baldizzi family from Sicily who moved into the tenement building in 1928.

Figure 2.1 The Baldizzi Apartment, 97 Orchard Street, New York City. Steve Brosnahan Collection of the Lower East Side Tenement Museum, New York City.

The Tenement House in Glasgow, owned by the National Trust for Scotland, tells the story of a home through the life and possessions of one resident, Agnes Toward, who lived there from 1911 to 1965. The tenement house was built in 1892, and had eight flats, each of which had a parlour, bedroom, kitchen, bathroom and hall. Miss Toward was a shorthand typist, and lived in one of these flats first with her widowed mother and then on her own until she went into hospital, where she died in 1975. The flat remained unoccupied for the ten years that Miss Toward was in hospital, and its subsequent owner sought to preserve it as it had been left, before selling it to the National Trust in 1982. The Tenement House 'is not a museum. What you see is a house and all its contents, from pictures in the parlour to pots of jam in the kitchen press, that once belonged to an actual person' (Ritchie 1997).

Box 2.2   HOUSING CAREERS

Housing careers are closely tied to, but distinct from, house biographies. Whilst house biographies illuminate the history of a particular dwelling through the lives of its various residents, housing careers chart the range of accommodation where an individual or a household lives over time. Through the quantitative and/or qualitative analysis of housing histories, this approach investigates residential mobility over the life course and can document the shifting experiences and meanings of home over both time and space. As Robert Murdie explains, '"Housing career" is a term used to describe the way in which households change their housing consumption as they move through the life cycle, or more generally, the life course' (2002: 425). The term does not necessarily imply a progressive movement from 'rental to ownership, or from multi-family to single-family or from small to large' housing, but rather records the series of dwellings a household occupies over time (Özüekren and Van Kempen 2002: 367). Özüekren and Van Kempen distinguish between large-scale longitudinal research, which follows households throughout their lives (as in Sweden: see Abramsson *et al.* 2002, and Magnusson and Özüekren 2002), and the wider use of smaller-scale, cross-sectional methods when respondents are asked about their housing histories.

As a recent special issue of *Housing Studies* (2002: 17, 3) shows, the analysis of housing careers can be particularly helpful in documenting and interpreting the housing inequalities experienced by minority ethnic groups. In his quantitative analysis of the housing careers of recent Polish and Somali migrants in Toronto, for example, Robert Murdie conducted a questionnaire survey of 60 respondents from each group who arrived between 1987 and 1994. Murdie structured the open and closed questions around a grid based on three homes: 'the first permanent residence, the one immediately before the current one, and the current residence' (428). Polish and Somali interviewers collected 'a brief summary of each move, some details about how the search was undertaken and the ease or difficulty of each search, and the objective and subjective characteristics of the housing they found and the neighbourhood in which the house is located' (428). Respondents were also asked about the extent to which their current accommodation felt like home:

> Polish respondents who were comparatively satisfied with their dwelling and viewed it as 'somewhat' or 'very much' a 'home' emphasized the importance of being with family, the presence of friendly neighbours and safety and security. For the Somali group, safety and security were primary considerations followed by proximity to relatives, friends and people from the same ethnic background.
>
> (439)

In contrast to such comparative studies, other researchers have investigated housing careers within particular minority ethnic groups. Bowes *et al.* (2002) stress the need to differentiate rather than homogenize minority ethnic groups, and conducted housing history interviews with British Pakistanis living in Bradford, Glasgow and Luton. They contacted people through 'snowballing' techniques, and aimed to interview people who lived across a range of housing tenure types who were in their 40s and thus 'of an age to have grown-up children who might be seeking new housing, and, possibly elderly parents, moving in, continuing to live with the family, or considering moving out' (387). Drawing on illustrative quotations from these qualitative interviews, Bowes *et al.* conclude that a wide range of factors – including gender, locality and class – cut across ethnicity in their influence on housing careers. As they conclude, 'poor housing conditions and housing disadvantage for Pakistanis are unlikely to be effectively alleviated by policies which focus only on ethnicity and race' (397).

Housing histories have also been important in the study of homelessness. Following Randall (1988), Jon May (2000a; also see May 2000b) uses the term 'homeless careers' in his biographical analysis of male, single, homeless hostel users in an English resort town. For May, such a biographical approach is 'essential if studies of homelessness are not to continue to deny homeless people an identity and agency beyond only their position as "homeless"' (2000a: 615). May interviewed his respondents about 'every accommodation and rough sleeping event, and the duration of those events, since a person had first left home' (2000a: 618), and constructed 'triple' biographies, which outlined changes in personal, employment and accommodation circumstances over time. Using pre-prepared sheets to record this information, May then drew a rough timeline, which was discussed and revised at the end of each interview. Analysing 'homeless careers' alongside personal and employment changes over time, May found that 'for the majority of single homeless people the experience of homelessness is neither singular nor long term but episodic, with each homeless episode interspersed with often extended periods in their own accommodation and with no increase in either the frequency or duration of homeless episodes over time' (2000a: 615).

early years of the Spruce Run Association, 'one of the nation's oldest organizations founded to advocate for battered women and to combat domestic violence' (www.umaine.edu/wic/both/FOHP). A script entitled 'The "Somebody Else" was us' was then developed by drawing on extracts from these oral histories, and is available via the website (see Warrington 2001; Meth 2003; and Chapter 3 for more on domestic violence).

In addition to such rich and diverse oral histories, many studies of home within and beyond geography draw on other interviews and ethnographic observations. Whilst ethnographic research on the home is usually conducted *in situ*, either on an episodic or a longer-term basis, interviews on the home are conducted beyond, as well as within, the domestic sphere. In her research on Filipino domestic workers in Vancouver, Canada, for example, Geraldine Pratt interviewed nanny agents as well as migrant workers. Such interviews show how the migration of Filipino domestic workers is racialized as well as gendered, whereby 'Filipinas are discursively constructed as housekeepers, with inferior intellects and educations relative to European nannies' (2004: 56), and are paid significantly less than their European counterparts as a result (see Chapter 5). In a different context, Ruth Fincher (2004) interviewed high-rise housing developers in Melbourne, Australia, to explore the ways in which domestic architecture and design are shaped by, and reproduce, particular inclusions and exclusions. Fincher argues that 'interpretations of gender are entwined in developers' views of the appropriate life course stage of the residents of inner city high-rise buildings' (327) and analyses the interview transcripts as narratives.

Other interviews are conducted within the home itself. In many cases, interviews are more than conversations and the research process incorporates a tour of the home (see Tolia-Kelly 2004a; Power 2005). The research questions addressed through such interviews and tours are wide-ranging, and include studies of suburban homes and families (Dowling 1998a and b); home-based work (Ahrentzen 1997; Oberhauser 1997); domestic material and visual cultures (Leslie and Reimer 2003; Rose 2003; Reimer and Leslie 2004; Tolia-Kelly 2004a and b; see p. 71); and experiences of home over the life course (including Kenyon 1999, on student homes). Such interviews have been particularly important in research on embodied experiences of home, as shown by research on domestic space, health and disability (Moss 1997; Imrie 2004a, b and 2005). In analysing and presenting home-based interviews, researchers typically follow the conventions of qualitative research. In Rob Imrie's research on disability and house design, for example, interviews usually lasted for two to three hours, were normally conducted within the subject's home, and spanned a range of themes that related to individuals' life histories and their experiences of home. The interviews were taped and transcribed, and transcripts were sent to each participant to check and, if necessary, amend. Imrie quotes extensively from these transcripts, and also includes a number of photographs that illustrate various changes to the home environment that have improved access and mobility.

Ethnographic research

In addition to interviews that are conducted within, as well as beyond, the home, ethnographic research is situated within the setting that is being studied. As Craig Gurney explains, 'Ethnography is a descriptive and analytical technique and perspective which has as its central endeavour *writing about a way of life*' (1997: 375), and usually includes detailed observation, in-depth interviews, and/or various other participatory methods. Ethnographic research 'involves an engagement of the researcher's senses and emotions. To engage a group's lived experience is to engage its full sensuality – the sights, sounds, smells, tastes, and tactile sensations that bring a way of life to life' (Herbert 2000: 552). It can be conducted in different places and cultures or close to home, and is particularly important in anthropology. Just as debates occur in geography about the location of the 'field', a number of anthropologists have written about conducting fieldwork at home. According to Irene Cieraad, for example, 'The anthropology of domestic space can become a native research paradise illustrating the exotic in the familiar' (1999a: 3; also see Messerschmidt 1981; Miller 2001). Reflecting on her research on the survival strategies of low-income women in Worcester, Massachusetts, Melissa Gilbert explains that research conducted 'at home' often involves learning about very different rather than familiar lives. For Gilbert, 'My lived experiences were so completely different from the women that I interviewed that I would not consider myself an "insider" despite the fact that I was doing my research on women who lived in the same city as I' (1994: 92).

Ethnography can insightfully highlight the multi-dimensionality and dynamism of ideas of home and home-making practices. In his research on the meaning of home for men and women living in working-class owner-occupier households in St George, East Bristol, Craig Gurney (1997) conducted a series of in-depth interviews with four key households, which he analyses as 'episodic ethnographies'. He describes such episodic ethnographies as 'the ways in which people make sense of home as a social construction. They provide the opportunity to reflect upon the significance of climactic events in personal biographies and the impact they have upon the meaning of home. Each climactic experience becomes a defining moment or a turning point' (376). As Gurney explains, ethnographic data does not claim to be representative:

> It is not the task of the ethnographer to generate research with validity for a wider universe, but to construct an internally coherent case study that

discovers the variables that have explanatory value in specific cultural contexts; in this case the ongoing social construction of four people's sense of home.

(376)

So, for example, Gurney identifies four episodes that have shaped Mr and Mrs Quinn's sense of home: uprooting, bereavement and marriage; parenthood and the meaning of home; the search for fulfilment; and the violent boyfriend. For Mr and Mrs Foley, in contrast, Gurney describes three episodes: 'doing things right'; bereavement and home as mausoleum; and post-natal depression and home as prison (for more on the ways in which old age, death and dying are excluded from idealized representations of home, see Hockey 1999).

Other ethnographic research involves living within particular households and making detailed observations about everyday life within the home. Arlie Hochschild, for example, was interested in both the practice – who did what – and perception – how people felt – of the gendered division of domestic labour. In her research based in California, she interviewed men and women, in their homes, about domestic tasks. But her interest in home-making practices also meant that she 'watched daily life', trying to become as 'unobtrusive as a family dog' (1989: 5). As she eloquently describes:

> I found myself waiting on the front doorstep as weary parents and hungry children tumbled out of the family car. I shopped with them, visited friends, watched television, ate with them, walked through parks, and came along when they dropped their children at daycare ... In their homes, I sat on the living room floor and drew pictures and played house with the children. I watched as parents gave them baths, read bedtime stories, and said goodnight. Most couples tried to bring me into the family scene, inviting me to eat with them and talk.

Although Hochschild does not critically reflect upon these experiences, others who have studied the home point out the importance of recognizing the power relations involved in research (see Dowling 2000). Researchers have an influence by simply being in someone's home and there are gendered and other dynamics (especially between couples) to interviews. Like all research, the safety of the researcher is a prime concern, and you should only visit respondents at home if you feel safe to do so and if you ensure that someone else knows where you are, how long you are likely to be there, and can contact you easily. Home-based interviews can potentially affect the researcher, often in unforeseen ways. In her research on the gender dimensions of low-income migration in South Sulawesi, Indonesia, Rachel Silvey lived in a dormitory with migrant women workers in the Makassar Industrial Region, or *Kawasan Industri Makassar* (KIMA) (2000a, 2000b; see Silvey 2003 for a discussion about her research methodology. Also see Mack 2004). She explains the importance, and challenges, of living with migrant women during her research:

> Living with migrants profoundly shaped my sense of factory workers' everyday lives, and my experiences in the dormitory have informed my interpretations and analyses following from fieldwork. The dormitory itself was constructed of plywood walls and corrugated aluminum siding. The walls were so thin that every noise in each room was audible in the adjacent rooms. Because factory workers held both day and night shifts, the noise was constant. Between four and six people slept, cooked, and lived in most of the small rooms. The sun heated up the aluminum siding, turning the building into an oven during the daylight hours, and leaving the rooms on the top floor unbearably hot until nightfall. The electricity blacked out at odd intervals, never providing a solid hour of light from the single bulb that hung in each room. The 53 residents in the building shared one water pump and one outhouse and bathing shed, and we lined up to use these facilities morning and night, each of us holding our pumped bucket of water as we waited in line to bathe.
>
> (2003: 97–8)

Silvey soon became ill living under these conditions. As she continues,

> Many days passed when I felt that the research was not progressing, and I often wondered if my discomfort, illness, and lack of privacy were wasted sacrifices. But it was through living with people in these conditions that my neighbours began to feel they could expose and explain to me their approaches to migration, factory work, and gender relations ... Actually experiencing, rather than simply observing, the exhaustion and everyday difficulties associated with living in the dormitory provided personal, experiential understanding of the implications of women's migration, urban lives, and factory work.
>
> (98)

WRITING HOME

In addition to life stories in text, other forms of writing are important sources in the study of home. In this section we focus on textual representations of the home in novels, magazines, and household guides (also see Research Box 1 by Alex Barley on the home in contemporary Indian novels). We explore imaginative geographies of home in both historical and contemporary texts, focusing on the ways in which accounts of home have been particularly important in writings by, for and about women. An important theme throughout this section is the way in which the home can be a site of creative containment and liberation (see Box 2.3).

Box 2.3   DOMESTIC CONTAINMENT AND LIBERATION: VIRGINIA WOOLF AND ALICE WALKER

In her classic work *A Room of One's Own*, published in 1928, Virginia Woolf wrote that 'a woman must have money and a room of her own if she is to write fiction' (Woolf 1945: 6; also see Shiach 2005). She asks why there was no female Shakespeare, and imagines that he had an equally gifted sister called Judith. Whereas William went to grammar school, married, moved to London, and worked as an actor, Judith remained at home:

> She was as adventurous, as imaginative, as agog to see the world as he was. But she was not sent to school. She had no chance of learning grammar and logic, let alone of reading Horace and Virgil. She picked up a book now and then, one of her brother's perhaps, and read a few pages. But then her parents came in and told her to mend the stockings or mind the stew and not moon about with books and papers.
>
> (48–9)

Woolf imagines that Judith Shakespeare ran away to London, where she tried to gain work as an actress. But this was seen as unsuitable work for a woman. She fell pregnant, and ' – who shall measure the heat and violence of the poet's heart when caught and tangled in a woman's body? – killed herself one winter's night' (50). Judith

---

provides a tragic counter to William's success as a writer. As Woolf concludes, 'it is unthinkable that any woman in Shakespeare's day should have had Shakespeare's genius. For genius like Shakespeare's is not born among labouring, uneducated, servile people' (50).

In an essay entitled 'In Search of our Mothers' Gardens,' Alice Walker (1984) draws on Woolf's essay to imagine the ways in which the creativity of black women in the past lived on despite centuries of oppression. Walker, the African-American author of novels such as *The Color Purple, The Temple of my Familiar* and *Possessing the Secret of Joy*, describes different forms of creative expression and the ways in which they are closely bound up with domestic life. She writes about seeing a quilt on display at the Smithsonian Institution by an anonymous Black woman in Alabama, a hundred years before, who used rags and scraps of cloth that were the only materials that she could afford, and expressed her creativity through the only medium that her position in society allowed her to use. Seeing this quilt makes Walker think of her own mother making quilts, towels and sheets, as well as raising five children and picking cotton all day. Walker describes her mother's garden as the site of her creativity and artistic expression: 'I notice that it is only when my mother is working in her flowers that she is radiant, almost to the point of being invisible – except as Creator: hand and eye. She is involved in work her soul must have. Ordering the universe in the image of her personal conception of Beauty' (241). Walker ends her essay by imagining a mother in Africa before she was sold and transported to the United States as a slave:

> perhaps in Africa over two hundred years ago, there was just such a mother; perhaps she painted vivid and daring decorations in oranges and yellows and greens on the walls of her hut; perhaps she sang ... sweetly over the compounds of her village; perhaps she wove the most stunning mats or told the most ingenious stories of all the village storytellers. Perhaps she herself was a poet.
>
> (243)

Both Woolf and Walker describe the ways in which the creativity of women in the past has been stifled. Woolf focuses on the material conditions that are necessary for women to write fiction, whilst

> Walker reflects on other forms of creative expression as well as writing. In each case, the home is seen as a site of both containment and liberation.

House and home in fiction

From domestic ritual to domestic transgression, and spanning memories and experiences of place and displacement, the home has been an important site and subject since novels were first written and read. According to Kathy Mezei and Chiara Briganti, the home and the novel 'have paradoxically constrained and liberated women's ways of knowing, writing and being' (2002: 844; also see Briganti and Mezei 2004). The novel has been interpreted as a form of domestic fiction, both in its content and settings, and in the material conditions of writing and reading. Mezei and Briganti write that 'novels and houses furnish a dwelling place – a spatial construct – that invites the exploration and expression of private and intimate relations and thoughts' (2002: 839). The private and intimate content of novels is often associated with writing by women because 'In writing from a domestic space of house, household, and family, women writers can create a position in the field of cultural production from which to value ordinary women's lives, the quotidian, the minute' (843).

The rise of the novel from the late eighteenth century has been associated with the growing separation of public and private spheres and the emergence of middle-class women as both readers and writers. Nancy Armstrong describes the rise of 'the domestic woman' as a major political event, and explains that British women began to write respectable fiction from the late eighteenth century, and became prominent novelists in the nineteenth century (Armstrong 1987: 3). In her 'political history of the novel', Armstrong argues that 'Domestic fiction mapped out a new domain of discourse as it invested common forms of social behaviour with the emotional values of women' (29). Similarly exploring the intertwined politics of the novel and the home, Elizabeth Langland traces the political significance of middle-class Victorian women within and beyond domestic life:

> Running the middle-class household, which by definition became 'middle class' in the possession of at least one servant, was an exercise in class management, a process both inscribed and exposed in the Victorian

novel. Although the nineteenth century novel presented the household as a moral haven secure from economic and political storms, alongside this figuration one may discern another process at work: the active management of class power. The novel, in sum, stages the conflict between the ideology of the domestic Angel in the House and its ideological Other (the Worker or Servant), exposing through women represented in fiction the mechanisms of middle-class control, including those mechanisms that were themselves fictions, stratagems of desire.

(Langland 1995: 8)

The domestic location and content of early novels was also important in American fiction. Ann Romines argues that housekeeping became an important subject in its own right in novels written from about the time of the Civil War in the 1850s. Describing housekeeping narratives in terms of 'the home plot', Romines explains that 'some of the best fiction by American women writers is dominated and shaped by the rhythms and stresses of domestic ritual' (1992: 9). For Romines, 'domestic rituals' are 'performed in a house, a constructed shelter, [and] derive meaning from the protection and confinement a house can provide. ... [A] domestic ritual can be a large, important household occasion, such as a family reunion or a home wedding, or it can be an ordinary household task such as serving a meal or sewing a seam. All such rituals help to preserve the shelter' (12). Domestic rituals were not only shaped by class and gender, but were also clearly racialized. As Romines observes, the home plots that revolved around such rituals were largely associated with white, middle-class women. Housekeeping for African-American women had a quite different history, which was inseparably bound up with slavery.

Just as home is open and porous (see Chapter 1), in the domestic setting of novels the very idea of home is closely tied to an idea of the foreign. Such 'foreignness' could be located within as well as beyond the nation, as shown by the way in which many novels by white, middle-class, American women from the 1850s represented Native Americans and African Americans as 'foreign' to both the domestic and national home. As Amy Kaplan explains, such novels

explore the breakdown of the boundaries between internal and external spaces, between the domestic and the foreign, as they struggle to renegotiate them. This struggle takes place not only within the home but also within the 'empire of the heart,' within the interior subjectivity of the heroine. Where the domestic novel turns inward into the private sphere of female subjectivity, we often find that subjectivity scripted by narratives of nation and empire.

(2002: 44; see Chapter 4 for further discussion)

The home continues to be an important site in contemporary fiction, particularly by women. In an edited collection entitled *Homemaking*, Catherine Wiley and Fiona Barnes introduce the 'politics and poetics of home' in writing by women in the second half of the twentieth century (Wiley and Barnes 1996; also see Pearlman 1996). Wiley and Barnes argue that 'Women write in order to negotiate the tensions between definitions of home as a material space and home as an ideal place' (xix), and show how women writing in different contexts and from different locations represent such tensions in a variety of ways. In her introduction to writings by working-class women, for example, Janet Zandy writes that

> Home is a good place to begin. Whether it is a tenement, a barrio, a ghetto, a neighborhood, the project, the block, the stoop, the backyard, the tenant farm, the corner, four walls, or hallowed ground, finding a place in the world where one can be *at home* is crucial. Home is literal: a place where you struggle together to survive; or a dream: 'a real home,' something just out of one's grasp; or a nightmare: a place to escape in order to survive as an individual. Home is an idea: an inner geography where the ache to belong finally quits, where there is no sense of 'otherness,' where there is, at last, a community.
>
> (Zandy 1990: 1, quoted in Wiley and Barnes 1996)

Different experiences and meanings of home are important in relation to sexuality as well as class, gender and 'race'. In an essay on constructions of home and family in Jeanette Winterson's *Oranges Are Not the Only Fruit* – an autobiographical, lesbian coming-of-age story set in an English evangelical home – and Jewelle Gomez's *The Gilda Stories* – a science-fiction fantasy, spanning two hundred years in the life of a lesbian vampire – Ellen Brinks and Lee Talley argue that 'many lesbian writers are reappropriating the notion of home' (Brinks and Talley 1996: 146; for more on lesbian geographies of home, see Johnston and Valentine 1995; Elwood 2000):

Poetically and politically fashioning domestic spaces that incorporate and welcome 'unnatural' passions, both novels expand our understandings of personal and social identity. Thus, while they write of contested practices that historically construct 'the family,' they also provide their heroines with a recovered sense of home and a kinship that is not based on blood ties but on choice. ... Both authors reject traditional and static representations of home with their emphasis on the inclusion of radically 'other' identities into the family. ... Their fantastic modes of representation suggest to the reader that what is unfamiliar may in fact be closest to home.

(168; domestic transgression and fantasy is also an important theme in other forms of writing, including gothic novels, ghost stories, science fiction and fantasy. See, for example, Ferguson Ellis 1989)

A number of postcolonial theorists have studied novels and other forms of writing to explore the ways in which the home on a domestic scale is inextricably bound to national and imperial geographies of home (including Rodriguez 1994). Edward Said (1993), for example, discusses *Mansfield Park* by Jane Austen (1814) in relation to wider relations of power and privilege beyond Britain, showing how the wealth of the Bertram family was rooted in slavery and plantation agriculture in the West Indies. Although Austen's novel is located in Britain, primarily within the domestic space of Mansfield Park, Said shows that this domestic, metropolitan setting can only be understood in relation to other, more distant, places. In other novels, these domestic and imperial connections come to be embodied within the home, most famously in the character of Bertha Mason, the Creole wife of Mr Rochester who is confined to the attic in Charlotte Brontë's novel *Jane Eyre* (also see the novel *Wide Sargasso Sea* by Jean Rhys (1957), which imagines Bertha Mason's early life. A play entitled *After Mrs Rochester* by Polly Teale (2003) focuses on Jean Rhys' life, and is set within various unhomely domestic spaces – symbolized by a locked room – inhabited by Rhys and Mason).

Other postcolonial critics have explored the unhomely displacements that destabilize 'the house of fiction' (Bhabha 1997: 445). Homi Bhabha, for example, writes that 'The home does not remain the domain of domestic life, nor does the world simply become its social or historical counterpart. The unhomely is the shock of recognition of the world in the home, the home in the world' (445). Drawing on *Beloved* by Toni Morrison and *My Son's Story* by Nadine Gordimer, and the homes within each novel that are haunted by slavery and apartheid, Bhabha argues that 'The unhomely moment relates to traumatic ambivalences of a personal, psychic history to the wider disjunctions of political existence' (448). Unhomely displacements are also an important theme within an 'immigrant genre' of literature (George 1996). For Rosemary Marangoly George, the homesickness often associated with this genre can be interpreted in two main ways: 'it could be a yearning for the authentic home (situated in the past or in the future) or it could be the recognition of the inauthenticity or the created aura of all homes. In the context of the immigrant novel it is the latter that usually prevails' (175; also see Peres Da Costa 1999, who discusses homesickness in the writings of Jamaica Kincaid, and see Rubenstein 2001, who analyses nostalgia and mourning in fiction by women. See Research Box 10 on p. 250 by Rachel Hughes on home and migration in Jamaica Kincaid's novels).

The home in magazines and household guides

We end this section by considering the imaginative and material geographies of home in magazines and household guides in both historical and contemporary contexts. Although such publications existed in countries such as Britain from the late eighteenth century, they became much more prominent in the nineteenth century, promoting a metropolitan view of middle-class feminine domesticity to readers both at home and across the empire (Armstrong 1987; Beetham 1996; Humble 2001; also see Chapter 4). Margaret Beetham identifies the *Englishwoman's Domestic Magazine*, published from 1852, as a watershed between exclusive ladies' magazines and more popular women's domestic journals, such as *Woman at Home* (1893–1920). Such magazines and a growing number of household guides helped to redefine middle-class domesticity and the feminine attributes on which it was seen to depend, and gave a new status to women at home who were depicted as competent managers with particular skills and knowledge. Yet the 'feminized spaces' of magazines and household guides were inherently unstable. At the same time as asserting their female, domestic, middle-class readership, such publications reproduced discourses of bourgeois femininity, based in large part on home-making, to which their readers were still aspiring. As Beetham explains, 'Throughout its history, the woman's magazine has defined its readers "as women." It has taken their gender as axiomatic. Yet that femininity is always represented in the magazines as fractured, not least because it is simultaneously assumed as given and as still to be achieved' (1). Household guides operated in a similar way, both asserting a feminized domesticity and instructing women on its achievement (see Box 2.4 for more on home economics).

Box 2.4   HOME ECONOMICS

The discipline of Home Economics attempts to formalize and to teach principles of domesticity. Although the term was not widely used until the early twentieth century, these attempts date back to the mid-1800s. HEARTH, the Home Economics Archive: Research, Tradition and History, is based at Cornell University in the United States, and provides an electronic archive of over 1,500 books and journals relating to home economics from 1850 to 1950 (http://hearth.library.cornell.edu/h/hearth/), as well as essays and bibliographies on subjects that include childcare, human development and family studies; food and nutrition; home management; housekeeping and etiquette; housing, furnishing and home equipment; hygiene; and retail and consumer studies. As the website explains,

> researchers in the field of women's history have been reevaluating home economics, developing an understanding of it as a profession that, although in some ways conservative in outlook, opened up opportunities for women and had a broad impact on American society. There was always a significant degree of disagreement among home economists, and among the legislators, policy makers, and educators who supported them, about what the field's mission should be. Some were focused on the home, while others were more concerned with the broader social environment. Some saw home economics as a vehicle for creating vocational and economic opportunities for girls and women and for educating boys and men about domestic skills, while others sought to enforce traditional models of sex roles and family life. However, even the most conservative models of home economics offered some women a path to careers as teachers and researchers.

Two particularly influential texts by American women that are available on this website are *Treatise on Domestic Economy* by Catharine Beecher (1841), which was later revised with her sister Harriet Beecher Stowe and published in 1869 as *The American Woman's Home* (2002); and *The Home: Its Work and Influence* by Charlotte Perkins Gilman (2002; first published 1903).

Studies of home magazines and household guides analyse the texts discursively, alert to the ways in which femininity and the domestic are represented. This often involves documenting and interpreting changes over time. In their research on the postwar home and the Australian housewife from 1940 to 1960, for example, Justine Lloyd and Lesley Johnson study a range of women's magazines (Johnson and Lloyd 2004; Lloyd and Johnson 2004; also see Hand and Shove 2004; Tasca 2004). Analysing the text and images that comprised mass-circulation magazines such as *Australian Women's Weekly* and *Australian Home Beautiful* (see Figure 2.2), Lloyd and Johnson argue that there was an erasure of

> the distinctions between decoration and planning, design and management, aesthetics and control. That is, the 1950s home increasingly articulated interior design *as* a form of household management. Whether budgeting for a new house or renovations, deciding on the kind of home the family would live in, or how it would be furnished, this discourse gave women a new capacity to shape (their part of) the world during the 1940s and 1950s.
>
> (2004: 255)

Lloyd and Johnson also point to a contradiction in postwar 'female domestic subjectivity':

> On the one hand, both state and market discourses suggested that women could sweep away the elements of traditional, particularly prewar, home designs that bound them to the home. On the other, popular magazines also placed a great deal of emphasis on the look of things and on looking itself, further inscribing women's identity within domestic space.
>
> (251)

The location and lived experiences of homes within these representations are variable. Deborah Leslie (1993), for example, has studied discourses of 'new traditionalism' in trade advertisements for the American editions of *Good Housekeeping* and *Family Circle* in the late 1980s and early 1990s. She analyses these advertisements in relation to the rise of new Right politics and broader economic and cultural shifts associated with post-Fordism, and identifies the contradictory nature of familial discourses whereby 'nostalgia for a more

Figure 2.2   Cover of *Australian Home Beautiful*, March 1946. Reproduced with permission of *Australian Home Beautiful* and courtesy of Mitchell Library, NSW, Justine Lloyd and University of Technology, Sydney, Digital Imagining Centre.

traditional family form [coexists] with the breakdown of the nuclear family' (691). These images depend upon and deploy 'images of rurality, small town America, and nature' (Leslie 1993: 704). More recently, Deborah Leslie and Suzanne Reimer have analysed home-design magazines such as *Elle Decoration*, *Wallpaper** and *Living* etc. Leslie and Reimer argue that new home-design magazines have been important in promoting the return to a modern aesthetic. In contrast to the 'new traditionalism' of magazines such as *Good Housekeeping* and *Family Circle*, new home-design magazines depict urban rather than rural homes and 'suggest that it is acceptable for women to have an interest in the more masculine realm of design' (304). As Leslie and Reimer explain,

These magazines articulate the notion that it is not oppressive for women to take an interest in their home, nor to view the home as an extension of fashion and their body. There is an implication that a flexible and modern style is liberating for women. At the same time, downplaying the gender specificity of home-making also enables magazines to target male readers. ... Unlike traditional home magazines that solicited a female consumer, new titles construct the consumer of modernism as a highly mobile young urban professional who might be male or female, straight or gay.

(304–5)

VISUAL AND MATERIAL CULTURES AT HOME

Household guides and magazines include images as well as texts, which alerts us to the importance of studying visual as well as textual depictions of home (for more on visual methodologies, see Rose 2001). In this section we explore the ways in which the home has been studied in relation to visual and material cultures. We begin by discussing the home as a site and subject in art, design and other media before turning to the use, display and meanings of objects within the home.

Domestic architecture and design

Architectural plans and interior designs provide a key visual source for analysing the home, as shown by research on the transnational domestic forms such as the bungalow and the high-rise (King 1986; Fincher 2004; Glover 2004); domestic architecture over time (Ravetz 1995; Sudjic and Beyerle 1999; Walker, 2002); style, taste and consumption (Sparke 1995; Madigan and Munro 1996; Bryden and Floyd 1999; Posonby 2003; Gram-Hanssen and Bech-Danielsen 2004); the gendered inclusions and exclusions of domestic architecture and interior design (Madigan and Munro 1991; Attfield and Kirkham 1995; Hayden 1996, 2002); and utopian visions of home, housing and community (see, for example, research on garden cities, including Hardy 2000; Pinder 2005). Whilst we explore these themes in greater detail in Chapter 3, we begin this section by introducing different methodological approaches to studying the home in relation to domestic architecture and design.

In a rich range of historical and contemporary contexts, many researchers have explored the ways in which domestic architecture and design are inscribed with meanings, values and beliefs that both reflect and reproduce ideas about gender, class, sexuality, family and nation. Louise Johnson's work on the 'display home' in Australia fits this theme (1993). For Johnson, house layout and the meanings ascribed to this layout through advertising, create home as a gendered and sexualized space. But there are clearly connections and disjunctures between idealized designs and the embodied practices of life at home (see, for example, our discussion of disability and the home in Chapter 3). A number of studies have explored this with reference to particular spaces within the home, notably the kitchen. In his work on British domestic modernities, for example, Mark Llewellyn (2004b) analyses the kitchen designs of the housing consultant Elizabeth Denby and the Modern architect Jane Drew in the first half of the twentieth century, and the ways in which they applied modernist principles of scientific efficiency in an attempt to rationalize the gendered spaces of domestic work (see also Freeman 2004; Hand and Shove 2004; van Chaudenberg and Heynen 2004; and two sets of themed papers on kitchens in *Gender, Place and Culture* (2006):13). As Llewellyn explains, 'Modern architects argued that the kitchen should be a machine for cooking in, but this small, modern, and efficient space, the heroic ideal of the architects, completely overlooked working-class social practice' (2004a: 240). At Kensal House, for example, kitchens were too small for a table, but 'just under a third of the residents ate [there] … on a regular basis in spite of its unsuitability for the purpose, perched up at the ironing board or at the shelf by the hatch' (2004a: 240). Such residents preferred to keep the living room for 'best'.

Relationships between domestic design and embodied practice have also been studied using other methods and sources. Inga Bryden (2004), for example, draws on oral history interviews in her study of the architectural and spiritual significance of the traditional courtyard house, or *haveli*, in Jaipur in northern India. Bryden explores the *haveli* as a gendered and inhabited domestic space, showing how the Hindu architectural principle of *Vastu Vidya* informs the everyday lives of families within a particular residence. As Bryden explains, 'Vastu symbolically and functionally connects the body of the individual with the spaces of the home and with a cosmological context' (36). It does so by defining the dimensions and orientation of the house and the direction of household activities. According to these principles, a *haveli* is designed on a nine-square grid that dictates internal divisions and the proportion of open and covered spaces. Bryden draws on oral-history discussions and interviews with mid- and high-caste families in their *havelis*. As she explains, 'these individualized accounts of the space of the home are, in part, viewed as a geography of intimacy where a sense of belonging, or feeling at home "in the world," is articulated in the form of story and oral history' (28). She explores the design and lived experiences of families within one particular *haveli*, and explains that 'The pattern of household activity in the *haveli* follows the movement of the sun, and the nature of the activity reflects the character associated with the direction' (36). So, for example, Pushpa, who lives in the *haveli* with her husband and children, explains that '"Direction does affect the householder's body and circulation of blood. For sleeping the head should be in the south-west because opposite energies attract" (the *Purusha*'s [cosmic man's] head is in the north-east)', and that '"The person who is cooking should always face east … then the food will taste better"' (36).

As well as influencing embodied practices within the home, domestic design itself is also an embodied practice, as shown by Renate Dohmen's work on the twice-daily production of *kolams*, or threshold designs, by women in Tamil Nadu, south India. Dohmen (2004) studies these acts of design through a performative reading of the relationships between the home and the world, and argues that the designs create a space of belonging not only for individual women and their families, but also for the community at large. As she writes, 'the designs, while referring to the private sphere of the domestic, are executed in the public sphere of the road, making the individual performative act of drawing a collective affair. … the presence of the designs clearly signal "home," "woman" and "wellbeing" to the wider community' (22).

The home and art

In addition to studies of domestic architecture and design, other work on domestic visual culture has focused on the home as both site and subject in art, photography and other media. In her feminist analysis of art, Griselda Pollock (1988) explores the domestic subjects of paintings by Berthe Morisot and Mary Cassatt, who were female Impressionists working in Paris in the late nineteenth and early twentieth centuries. Unlike the wide vistas depicted in the paintings by many of their male counterparts, their domestic settings were depicted from a much more proximate and embodied perspective (also see Cieraad 1999b for a study of windows in paintings of seventeenth-century Dutch interiors). Although the home has been an important subject in certain artistic traditions, other traditions are profoundly unhomely. According to Christopher Reed, the home has long been suppressed in art, particularly in the work of the modernist avant-garde:

From the Victorian drawing-room with its étagères full of trinkets to the twentieth-century tract house with its mass-produced paintings, the home has been positioned as the antipode to high art. Ultimately, in the eyes of the avant-garde, being undomestic came to serve as a guarantee of being art.

(Reed 1996: 7; also see Reed 2002)

And yet, in recent years, Haar and Reed explain that 'In contrast to the ambivalent – even antagonistic – relationship between domesticity and modernism, the postmodern era has witnessed a kind of homecoming in high culture, as artists and designers have (re)turned their attention to domesticity' (1996: 253). This 'homecoming' raises important questions about the home as a site, as well as a subject, for artistic practice, performance and display (see Box 2.5 on Rachel Whiteread's *House* and Michael Landy's *Semi-Detached*). A number of recent exhibitions have focused on domestic subjects in art, including – in London alone in the last two years – 'Home and Garden' (Geffrye Museum, 2003), 'Below Stairs: 400 years of Servants' Portraits' (National Portrait Gallery, 2004), and 'Art of the Garden' (Tate Britain, London, 2004). Clearly, homes and gardens have been, and continue to be, important subjects in visual art. But the home as a location for art remains more contentious. As Colin Painter explains, 'If the domestic has never been rejected as a subject, antipathy toward the home as the ultimate, and "permanent", destination for art has been clear and persists more clearly' (2002a: 1).

Box 2.5   *HOUSE* AND *SEMI-DETACHED*

The built form of the house has been used for artistic purposes in a variety of ways. Perhaps most famously, Rachel Whiteread's *House* (Figure 2.3) was commissioned by Artangel, completed in 1993 and demolished in 1994 (www.artangel.co.uk; for more information and photographs of *House* visit www.artistsineastlondon.org, and see Lingwood 1995). *House* was a cast of an 1870s terraced house in Grove Road, Bow, East London, rendering a private space as a public sculpture by turning the inside of the home onto the surface of a concrete cast. As Doreen Massey writes, *House* was disruptive and unsettling in a number of ways. First, 'it made present something which was absent; it was the space of a house no longer there' (1995b: 36). Second, 'it turned the space inside out. The private was opened to public view. The little intimacies were exposed: the print of the flex running down the wall to the light switch seeming so personal, so vulnerable now' (36). Third, *House* solidified 'the volume that had once been the interior of the house: the living space, the space of life' (36). According to James Lingwood, Co-Director of Artangel,

> Like many public sculptures and memorials, 'House' is a cast. But unlike the bronzes which commemorate triumphs and tragedies, great men and heroic deeds, this new work commemorates memory itself through the commonplace of home. Whiteread's in situ work transforms the space of the private and domestic into the public – a mute memorial to the spaces we have lived in, to everyday existence and the importance of home.

> (quoted by www.artistsineastlondon.org)

And yet, Whiteread's work was called *House* rather than *Home* (see Chapter 3). For Massey, the naming of the sculpture reinforced its disruptive effects by defamiliarizing 'the normal, comfortable mythologizing of "home"', and by revealing its complexity whereby 'its meaning always has to be interpreted; that there was never any simple "authenticity"; that the meaning(s) of home are always open to contestation' (42).

House was the focus of considerable media attention and public debate. On the same evening in November 1993 that Rachel Whiteread was awarded the prestigious Turner Prize, Bow Neighbourhood demanded the immediate demolition of *House*. Although *House* was always intended to be temporary, Artangel had been attempting to negotiate an extension because delays had meant that there were only six days between its completion and the scheduled date of its demolition. After the matter was discussed in parliament, Bow Neighbourhood agreed to an extension, and *House* was not demolished until January 1994 (Lingwood 1995).

Figure 2.3 House, Rachel Whiteread, 1993. Commissioned and produced by Artangel; Photograph by Stephen White. Reproduced by permission: Artangel.

Whilst *House* was located in situ, on the site of the house that it inverted, Michael Landy's *Semi-Detached* is a full-size replica of his family home in Ilford, Essex, but located within the Duveen Galleries at Tate Britain in London from May to December 2004. Figure 2.4 is an artwork by Michael Landy depicting his parents standing in front of *Semi-Detached*. Best known for his earlier work *Break Down* (2001), when Landy shredded all of his possessions over a two-week period, *Semi-Detached* was a reconstructive work that centred on his father's life at home since an industrial accident in 1977 ended his working life as a tunnel miner (Landy 2004). Through the physical reconstruction of his parents' house, and the use of video, Landy explores the domestic confinement of his father, John. As Judith Nesbitt writes, 'The installation presents the front and rear elevations of the Landy's semi-detached suburban house, reproduced to scale and using real bricks and mortar. Utterly familiar, it has the characteristic features of millions of British houses, "modernised" with pebble-dashing, UPVC windows and doors, satellite dishes and kitchen extensions' (2004: 15). Inside the house, one video screen projects images of the bedroom, dining room and garden shed, whilst another focuses on the contents of a shelf in John Landy's bedroom. On the rear façade of the house, another video shows a sequence of images from DIY manuals, instruction leaflets and home-improvement magazines collected by John Landy both before and after his accident: 'Photographs and line-drawings of optimistic young couples and growing young families, hell bent in pursuit of the modernity, pureness and newness that was all the rage, alternate with illustrations of how to deal with blocked guttering, eroded surfaces, skinned knuckles, clogged drains' (Burn 2004). *Semi-Detached* presents an embodied home space, shaped by the life of Landy's father.

At first glance, *Semi-Detached* represents 'a reaffirmation of identity through the reassuringly familiar, if banal, qualities of the

Figure 2.4   Semi-Detached, John and Ethel Landy, 2004, Michael Landy. Copyright Michael Landy; reproduced courtesy of Thomas Dane, London.

everyday settings and objects that constitute home and "homeliness"' (Melhuish 2005: 118). But the sounds and images projected
behind and onto the suburban façade, and the very process of transforming 'the material culture of British suburbia into a subject of art'
(121), render the apparent homeliness of *Semi-Detached* unfamiliar
and strange. As Clare Melhuish writes, *Semi-Detached* 'takes the
viewer away from the realm of the home, and into another, that of the
tangibly "unhomely," where all that should seem familiar and pregnant with significance, becomes distant, strangely foreign, and open
to fresh understandings' (121).

In an attempt to redress this antipathy, David Halle argues that the home is
just as important in understanding the meaning of art for its audience 'as the
mode of production or the public world of museums, galleries, and critics'
(1993: 193; also see Box 2.6 on the 'At Home with Art' project). Halle
focused on four areas in the New York region to provide a cross-section of
neighbourhoods by class, ethnicity and urban/suburban location. Within
each area, Halle selected a random sample of houses, and then sent a letter to
each house before visiting with a research assistant. In addition to interviewing adult residents, Halle was also taken on tours of the home, took
photographs of the interiors and drew floor plans. Focusing his study on landscapes and family photographs, abstract art and 'primitive' art among the
upper middle class, and religious iconography among the Catholic working
class, Halle not only quotes from interviews but also employs a range of statistical techniques in his analysis. So, for example, he investigates landscape
pictures as markers of status and class differences by using multiple regression
analysis to control other possible causal factors such as religion, age, and the
number of landscape paintings in each house. Although certain common
features emerged – only 2 out of 349 landscape paintings depicted scenes that
were turbulent rather than tranquil, for example, and 71 per cent of historical
landscapes were populated, compared to only 11 per cent of contemporary
landscapes (72) – others were differentiated by class. As Halle explains, 'the
identity and prestige of the artist; the frequency with which foreign societies
in general are depicted; the frequency with which certain foreign societies –
Japan, Britain, and France – are depicted; and the frequency of "historical"
landscapes that depict the past' (80) were much more notable features of

landscape paintings in upper-middle-class, rather than lower-middle- and working-class, homes. Both Halle and the 'At Home with Art' project (Box 2.6) point to the importance of the home as a location for art. Unlike museums, as Halle explains, 'in the house the main audience – the residents – are also the ones responsible for the presence and arrangement of the art and cultural items. This is why the mode of dwelling provides a locus for under-standing new, and unexpected, ways in which art signifies in the modern world' (200).

Box 2.6   'AT HOME WITH ART'

'At Home with Art' was a project organized by Colin Painter that involved nine sculptors making objects for the home, which were then mass-produced and sold across Britain in Homebase stores. The project also involved an associated exhibition at the Tate Gallery (now Tate Britain) in London in 1999/2000, which then toured Britain for two years, the production of a BBC documentary entitled *Home is Where the Art Is*, and the publication of a collection of essays (Painter 2002b). The project had three main aims:

- to reconsider the home environment as a context for contemporary fine art and review ideas and values associated with the domestic.
- to explore the possibility of contemporary art reaching a wider public through work being made specifically for the home.
- to make available to the mass market objects made by contempo-rary artists.

(Painter 2002c: 7; see also Painter 1999)

The emphasis of the project, Painter explains, 'was on positive participation in domestic life as distinct from critique. The brief required that each artist began by visiting a household to observe the way in which visual objects and images were displayed and under-stood there. This was to ensure a genuine focus on domestic reali-ties' (2002c: 8). The sculptors produced six functional objects and three 'pure' sculptures, which included a sound sculpture, a garden trowel and fork, a peg, and a shower curtain, ranging in price from

---

£6.99 to £56.99 (see Deacon *et al.* 2002, for the reflections of three participating sculptors on the project). Drawing on eighteen in-depth interviews with people who purchased at least one of the items produced as part of the project, Rebecca Leach (2002) explores their meanings and display, and the different ways in which people 'curate' their homes. As she writes,

> The findings indicate the embedded nature of things and cultural objects in people's lives: things in homes cannot be separated easily from families, financial circumstances, individual and broader senses of taste, cultural ideas and myths about how homes should be. There is one instance ... in which the AHWA objects are taken into the home in an explicitly aesthetically discriminating manner – as art objects carefully chosen because of their place in the art world. Such respondents treat their homes almost entirely as galleries, curating the space and carefully managing its displaying qualities. For most respondents, however, the judgements are made more haphazardly, reflecting the multiple demands that homes place on us.
>
> (153)

Performance art and the home

Performance artists also provide creative and diverse examples of the home as a site, as well as a subject, of visual culture, and raise important questions not only about the performativity of home and domesticity, but also about the relationships between art and the home. For example, in 'Kitchen Show', first performed in 1991, Bobby Baker performs in her own and others' kitchens and 'inflects the multiplicity and heroism of work in the kitchen with flights of imagination' (Floyd 2004: 65; also visit Bobby Baker's website on www.bobbybakersdailylife.com. This website also includes information about 'HouseWorkHouse', which is a project with two older-people's groups in London. Household objects provided a starting point for discussion and storytelling in workshops, and are positioned in different rooms that you can visit via a web installation). As Janet Floyd explains, Baker 'cuts loose from the idea of the kitchen as private but altogether predictable space. She acts out a set

of behaviours that give a surreal twist to the tasks and activities that are continually presented to us in popular culture, and to the familiar association
between the mundanity of women's domestic lives and flights of feminine
fantasy' (65). In so doing, as Floyd and other critics observe,

> this is a performance that breaks down what might seem to be an imper-
> meable boundary between artist and housewife, between the spaces
> where 'art' is practised and the abject spaces of the home. Equally, this is
> a kitchen text that, as well as insisting on the richness of domestic experi-
> ence, and specifically of the kitchen, as a source for creativity, also seeks
> to ridicule the pretensions of the art world.
>
> (65)

Like Bobby Baker, other performance artists have also used the space of the
home as integral to their art. In London, for example, a house in Camberwell was
opened as home in 1998: 'a radical new performance art space' based within a family
home. As Laura Godfrey Isaacs – artist, curator and resident of home – explains,

> Rather than clearing the rooms so they operate as a white space, artists
> make use of the living environment of the house to inform, inspire and
> provide a context for their work ... Over fifty performance events have
> taken place at the house and almost every room has been used. The
> venue provides a unique, intimate and informal viewing experience for
> the audience. There is an absence of boundaries and protocols of gallery
> or theatre in this specific domestic environment, allowing a close interac-
> tion between audience and performer.
>
> (www.lgihome.co.uk)

The domestic space of the performances, and the close interaction between
visitors and artists, informs the salon series and other performances that take
place at home. From January to March 2005, the 'One to one' salon series
involved individual domestic encounters and interactions between artists and
visitors. At 'A Day of Domestic Bliss', for example, home was open all day for 'a
choice of three separate interactions; have your dirty laundry washed, your
ironing done or enjoy a cup of tea or coffee with an artist at a classic coffee
morning' (www.lgihome.co.uk).

Family photographs

Alongside the 'homecoming' of visual and performance art, other media such as film, television, video games, the internet and family photographs have been particularly significant in studying domestic visual cultures (see, for example, Box 2.7 on the home in film; Morley 2000 and Spigel 2001 on television, home and family; McNamee 1998 and Flanagan 2003 on video games; and Holloway and Valentine 2001 on children's use of the internet at home). Developing the argument about the significance of the home as both site and subject of visual culture, we want to focus on family photographs. In both historical and contemporary contexts, a wide range of research has analysed family photographs in relation to imperialism, memory, autobiography, and gendered and racialized domestic spaces (including Spence and Holland 1991; Kuhn 1995; Curtis 1998; Blunt 2003b; Chambers 2003; Rose 2003, 2004).

Box 2.7   THE HOME IN FILM

The home is an important location in many films across a range of genres (also see our discussion of television in Box 3.1). Rather than view the home solely as a background to a wide range of films, many critics explore the ways in which it plays a more central and active role. Describing the suburbs as 'a cinematic fixation' in late-twentieth-century American film, for example, Douglas Muzzio and Thomas Halper distinguish between 'suburban set' and 'suburban centred' feature films (2002: 544; 547) that revolve to a large extent around home and family life. Whilst the former term refers to those films that could have been set elsewhere, 'sometimes suburbia is so essential to a film's nature that it could not take place elsewhere without being fundamentally altered' (547). Amongst a wide range of examples of such 'suburban centred' films, Muzzio and Halper discuss *American Beauty* (1999), *The Truman Show* (1998) and *Pleasantville* (1998). Unlike early television sitcoms 'that celebrated the goodness, wholesomeness, and fun of the idealized suburb, glorifying such bourgeois virtues as honesty, courtesy, cheerfulness, obedience, and neatness' (548), more recent films subvert and/or satirize this ideal by depicting American suburban homes and families as sites of dysfunction, duplicity, shame and pain (also see Mortimer 2000, who describes the 'suburban grotesque' of recent Australian feature films, and Goldsmith 1999. For a hyperreal representation of postwar American suburban home and family life, see *Far From Heaven* (2002)).

Other critics have explored the relationships between house and home in film. Through his focus on the American film *Panic Room* (2001), for example, Peter King (2004) explores issues of privacy, security and anxiety and argues that film criticism can be an important method for housing research. Jane M. Jacobs analyses the film *Floating Life* (1995), tracing 'the interface between the emotional experience of feeling at home and the architectural materiality of the house as formed through the drama of mobility' (2004: 165). In this film, the migration of a Chinese family from Hong Kong to Australia and Germany is conveyed through '[t]he idea of the house.' As Jacobs explains,

> The many houses of *Floating Life* are not simply the privileged *context* of action in this film. Interactions within and with the houses are positioned as animated components in the daily lives of their inhabitants: setting off thoughts, dictating action, mediating well-being, expressing identity.
>
> (171; see Chapter 5 for more on diasporic homes)

Other genres of film also explore the relationships between house and home. Ceridwen Spark (1999; 2003), for example, contrasts two documentary films made about the Block in Redfern, which is a small area of Aboriginal-owned and inhabited terrace housing in central Sydney. Established in the 1970s by an Aboriginal housing corporation with funding provided by the Australian government, the Block raises important questions about what home means in colonial settler societies such as Australia (Anderson 1999; see Chapter 4 for further discussion). Redfern is configured in unhomely ways by the popular media. Newspapers and television reports regularly depict it as a ghetto, as an uninhabitable place, and as a place that is out of control. But, in contrast, Aboriginal residents talk about Redfern as home and construct a sense of belonging, whereby the Block becomes a site of resistance and a place from which to draw strength and to negotiate power relations. As Spark explains: 'Non-Aboriginal attempts to deny the existence of homeplace through the designation of The Block as unlivable ghetto, or drug-infested hell-hole are subverted by the continuing sense of home produced in and by the bodily/placial practices of Aboriginal people' (2003: 60).

As part of her research, Spark studies two documentary films and, in particular, their focus on Aboriginal homes in Redfern. One film was made by the Australian Broadcasting Corporation (ABC) and broadcast in 1997, whilst the other was made by the Special Broadcasting Service's Indigenous Cultural Affairs Magazine (ICAM) programme and broadcast two years later. Rather than analyse both films in their entirety, Spark focuses instead on particular sequences within them, and contrasts the use of domestic settings in interviews with Aboriginal women. As she explains,

> The domestic focus of the sequences promotes a particular kind of viewing relation in which the audience is placed in close proximity to personal matters. Associated with the private, and thus with the supposedly unmediated world of the emotions, the internal sphere tends to be constructed as more 'authentic' and 'feminine' than the public realm.
>
> (2003: 37)

Whilst the ABC film interviews women within the home, most of the interviews with women in the ICAM film are conducted in more public places such as the street and parks. According to Spark, the ICAM film seeks to represent Redfern – the suburb beyond the domestic sphere – as a homeplace, whereas the ABC film 'reproduces the notion that questions about the Block are best answered by women discussing housing conditions from their lounge rooms or rat-infested kitchens' (38). As a result, Spark continues, 'dwelling is represented through images of failed domesticity rather than the typically more felicitous local race- and culture-based forms of belonging that exist in and around the streets of Redfern' (38).

Whilst most studies of family photographs concentrate on those taken by family members and displayed within the home, some explore those taken by other photographers and reproduced for a range of purposes beyond the home. In his study of photographs of rural homes in the American Great Depression in the 1930s, for example, James Curtis (1998) analyses some of the 80,000 photographs taken under the auspices of the Farm Security Administration. The FSA photographic project used 'the powerful symbols of home and family to generate support for government relief efforts' (275). But, as Curtis explains, the celebratory portrait of the rural poor as 'needy but not helpless victims of the depression' (275) excluded African Americans, who made up a significant proportion of the rural labour force: 'Black share-croppers and migrant workers appeared in the FSA file with great regularity – but as negative not positive symbols. African American families, homes, and material possessions fell under the relentless gaze of imagemakers who stripped away the marks of civilization in a quest for the primitive' (276).

Within the home, family photographs, and, in particular, their compilation into family albums, 'represent ideas about spatial identity and belonging', and the conventions that their compilers use 'to domesticate space' (Chambers 2003: 96; also see Chambers 1997). Deborah Chambers analyses photographs from her own family albums as well as oral history interviews with ten, Anglo-Celtic, white women who live in Western Sydney about the family albums that they compiled in their twenties. Chambers argues that family albums provide narratives that structure meanings over time, document particular events, and can act as 'visual family trees' (100). According to Chambers,

> when the extended family unit was being replaced by a smaller, privatized, geographically and socially mobile family, amateur photography and the family album as a form of display came to the rescue to mythologize a fixed version of it. Family photography records and reinvents something that no longer exists.
>
> (101)

Family photographs are, of course, on display in the home as well as compiled into albums. In New York, for example, David Halle (1993) found that family photographs are often clustered together in free-standing arrangements, making it 'simple to subtract, add, and regroup photos' (115). For Halle, family pictures are closely connected to social and material life, especially to that of house and neighborhood. In family photos we can trace the decline of formality in the house, the emergence of new conceptions of the purpose of life outside the workplace, the instability of modern marriage, the demise of boarding and lodging, and the weakening of financial obligations toward older, living kin.

(116)

Gillian Rose (2003; 2004) has also studied the meanings attached to family photographs, but did so by interviewing fourteen white, middle-class, married women with young children who live in two towns in the south-east of England. Her study of the relationship between domestic space and family photographs is thus 'quite specific in terms of both the "family" and the "domestic"' (2003: 6). Rose explains that 'family photographs are an important means through which a "home" is made from a house. But family photography also stretches "integration" beyond the house' (2003: 7) (see Chapter 3 for further discussion).

Divya Tolia-Kelly has also studied the domestic display and meanings of family photographs, alongside images of landscape, religious iconography and particular home possessions (2004a and b) in the homes of South Asian women living in London. Drawing on interviews, tours of the home and discussions about particularly significant objects and images, Tolia-Kelly shows that geographies of home extend far beyond the household, charting routes and connections between the past and the present and across diasporic space (see Chapter 5). As she argues, 'These visual cultures operate beyond the mode of the visual, incorporating embodied memories of past landscapes and relationships with pre-migratory lives in colonial territories' (2004a: 685). Tolia-Kelly has also worked with the artist Melanie Carvalho to depict ideal 'landscapes of home' for the women that she interviewed (Anderson et al. 2001; see Box 2.8 on imagining home).

Box 2.8   IMAGINING HOME

The visual arts not only provide an important source for understanding home, but can also be used as a research technique by asking people to visually depict past, present and/or ideal homes.

The artist Melanie Carvalho, for example, has talked to nearly 100 people about their ideas of home, asking them to describe in a piece of writing and/or a simple sketch 'the landscape that best represents your idea of home' (Anderson *et al.* 2001: 113; see also Tolia-Kelly 2006). From these responses, Carvalho then paints a landscape, seeking to be 'as objective as possible, and respond as technically as possible, without elaboration' (115). As part of this project, Carvalho collaborated with the geographer Divya Tolia-Kelly, and participated in workshop discussions with South Asian women in London. She subsequently painted seventeen landscapes of home, which were exhibited in 2000 and 2001, and can be viewed at www.geog.ucl.ac.uk/~dtkelly/publications_media.htm. Tolia-Kelly's research was on the ways in which British Asian women related to landscape and involved three workshop sessions with women recruited from two British Asian women's groups in London, followed by interviews at home, tours of the home, and discussions about visually important objects and texts in the home. As she explains, 'I had decided to do one session in which women brought in objects that were important to them, such as photographs, videos and so on, and one biography session. But I wanted a session where they would be creating something, to give them some freedom' (116). Carvalho led this session, asking the women to write about and/or draw their landscapes of home. In contrast to other participants in her project, Carvalho found clear differences both in the content and style of visual representation:

> Other people gave me more abstract and conceptual responses. The women in the workshops described very specific places. I don't know whether that's to do with the fact that they have been disconnected from where they originally came from – but it seems that the descriptions I got from people from Britain who have always lived here are much more abstract. A lot of the women described a childhood home, a place where they'd actually been. Many of them were worried that their drawings would look childish, and they wanted me to make them more picturesque. But they didn't look childish to me: they looked beautiful. They weren't a western mode of drawing, because they were

quite a flat perspective, there was a lot more pattern than there is in the western tradition of landscape. I tried to incorporate that kind of pattern into my paintings.

(118)

In a fascinating study of people's feelings about home, Claire Cooper Marcus (1995) also uses visual techniques as an important and innovative part of her research. Interviewing more than 60 people in their homes, mostly in the San Francisco Bay Area, Cooper Marcus explored their strong emotions – both positive and negative – about home. As a starting point, Cooper Marcus asked participants to

put down [their] feelings about home in a picture; I supplied a large pad of paper, crayons, and felt pens. If they objected with 'Oh – I can't draw,' I reassured them that this was not a test in drawing, but rather an opportunity for them to focus on their feelings without speaking'.

(8)

Cooper Marcus found that people depicted home in a wide range of ways: 'Some people did childlike house diagrams with words or colors indicative of feelings. Others produced mandala-like symbols, semiabstract images, or artistic renderings' (8). While her interviewees were drawing their pictures, Cooper Marcus spent about twenty minutes looking at the rest of the house or apartment, taking photographs, and making notes. On her return, she asked the interviewee to describe their picture, and then to participate in role-play. As she explains,

I would place the picture on a cushion or a chair about four feet away and would ask them to speak to the drawing as if it were their house, starting with the words, 'House – the way I feel about you is ...'. At an appropriate moment, I would ask them to switch places with the house, to move to the other chair and speak back to themselves as if they were the house. In this way, I facilitated a dialogue between person and house, which often became quite emotional, sometimes generated

> laughter, and occasionally brought forth statements begin-
> ning, 'Oh, my God ... ,' as some profound insight came into
> consciousness.
>
> (8–9)
>
> Through this process, Cooper Marcus was able to explore both
> conscious and unconscious feelings that people have about their

Domestic material cultures

Studying material objects as well as images within the home, Tolia-Kelly's research is informed by a concern to rematerialize geography (see, for example, special issues of *Geoforum* (2004) on material geographies and *Social and Cultural Geographies* (2003) on 'culture matters'), and an engagement with the interdisciplinary literature on material cultures. The home is a rich site and subject for research on material cultures (see Research Box 2 by Caitlin DeSilvey on domestic archaeologies and material cultures, and also see Research Box 6 by Sarah Cheang on Chinese material culture in British homes in the late nineteenth and early twentieth centuries). In both historical and contemporary contexts, domestic material cultures have been studied in rela-tion to particular objects within the home, their changing use and meanings over time, and the broader politics and practices of domestic consumption and display (see, for example, Csikszentmihalyi and Rochberg-Halton 1981; Ferris Motz and Browne 1988; Thompson 1998; Parr 1999; Attfield 2000; Miller 2001; Leslie and Reimer 2003; Pink 2004; Reimer and Leslie 2004). Such studies often explore not only material cultures within the home, but also the ways in which such material cultures should be understood in relation to the wider world, whether in terms of commodity chains of production, retail and consumption, memories of migration and displacement, or broader processes of economic and cultural globalization (see Chapter 5 for further discussion). As Daniel Miller (2001: 1) explains, 'It is the material culture within our home that appears as both our appropriation of the larger world and often as the representation of that world within our private domain.'

Research on domestic material cultures involves a wide range of methods, including archival analysis, object biographies, interviews and ethnographic research (for a different approach to studying the materiality of home, which is inspired by science studies and explores the importance of non-human agency, see Hitchings 2004, and Box 2.9 on the entanglements of nature and culture at home). In her discussion of the furnishing of slave quarters at Colonial Williamsburg in the United States, for example, Martha Katz-Hyman (1998) provides a detailed account of the historical sources used to inform decisions about what objects to include, where they might be placed, and how they might be used. Colonial Williamsburg is the world's largest living history museum. Extending over more than 300 acres in Virginia, and consisting of hundreds of restored and reconstructed buildings, it represents 'the restored eighteenth-century capital city of Britain's largest, wealthiest, and most populous outpost of empire in the New World' (from the website of Colonial Williamsburg: www.history.org). As part of this site, the four wooden cabins that comprise the slave quarter at Carter's Grove represent life for rural slaves in eighteenth-century Virginia (see Figure 2.5). Katz-Hyman describes the research involved in deciding how to furnish these slave quarters, prompted by the following questions:

> Did slaves have only the bare essentials for living, or did they have some of the amenities that were usually obtained by whites throughout the colony? Other than being given goods by their masters, were there other

Figure 2.5   Slave quarters at Carter's Grove, Colonial Williamsburg. Reproduced by permission Colonial Williamsburg Foundation.

ways in which slaves acquired their material goods? What were these ways? What did slaves do with these goods? How did they arrange the possessions they did have? After nearly two hundred years of Virginia slavery, did blacks maintain any African traditions, and if so, how were such traditions reflected in slaves' material goods?

(197)

As Katz-Hyman writes, there was a surprising amount of evidence from sources that included probate inventories, letters and diaries, accounts by travellers, account books of both merchants and craftsmen, advertisements for runaways, archaeological recovery, and accounts by eighteenth-century travellers to Africa that 'offer insights into cultural traditions that may have persisted in America' (205). But, although such sources helped to reveal how slaves obtained their goods and what they were, harder to answer were questions about their use, placement, and appearance: 'How did people use these things? Where did people sit? Where did they store their possessions? How "new" or "old" did these things look? How cluttered or neat were these spaces? Did these spaces look the same as spaces occupied by poor whites? In short, how did these spaces look two hundred years ago?' (209). The different scenarios presented in the slave quarters at Carter's Grove thus reflect both historical research and conjecture.

Box 2.9   HOME, NATURE AND CULTURE

Situated within broader debates about the entanglements of nature and culture (including Castree 2005; Whatmore 2002), a growing range of research focuses on human and non-human cohabitation, particularly in relation to the philosophy and ethics of domestication and within the space of the private garden. Maria Kaika (2004), for example, explores the relationships between nature and the home. Through a focus on water, she investigates 'the material versus the ideologically constructed boundaries between "the natural" and "the domestic" space', and examines the ways in which nature became 'scripted as "the other" to the private space of the bourgeois home in western societies' (266). Kaika argues that 'the dwelling places of modernity are *hosts* of the uncanny in their very structure' (281), as shown by the simultaneous need and denial of socio-natural processes within the home. During moments of crisis, she argues, the networks, pipes and other material manifestations of the connections between 'natural' and 'domestic' spaces surface as 'the domestic uncanny' (266). According to Kaika, 'Exploring the uncanny materiality of "the other" in the form of the invisible metabolized nature or technology networks points at the social construction of the separation between the natural and the social, the private and the public' (283). Homes are not always 'unnatural'. The website of the House Rabbit Society (www.rabbit.org) provides fascinating material on transforming houses into rabbit habitat (see also Smith 2003).

Other research on the domestic entanglements of nature and culture has concentrated on the space of the garden, the embodied practices of gardening, and horticultural agency. Recent research has investigated the garden in art (Postle *et al.* 2004); landscape and transculturation in Japanese gardens in Edwardian Britain (Tachibana *et al.* 2004); the relationships between property, boundaries, and public and private gardening (Blomley 2004a and b); and the meanings that people attach to gardens and to gardening (Brook 2003; Bhatti and Church 2004). Other work on gardens and gardening has been inspired by actor network theory. Russell Hitchings (2003), for example, explores the relationships between eight gardeners and the plants in their garden, studying 'how human and non-human actors worked together in the process of creating a garden and how these processes informed the human conceptions of these gardens' (102). This is part of a larger research project on 'the changing material agencies and entities that find their ways into the domestic gardens of London', which has focused on four specific contexts: 'the London garden centre, the garden designer's studio, the designed garden and the experienced gardener's garden' (Hitchings and Jones 2004: 8). As this work shows, human and non-human agency shapes the space of the domestic garden, providing important and challenging insights for studying various cohabitations at home and beyond.

Research on domestic material cultures in the more recent past and in the present often includes interviews and ethnographic observation alongside archival and visual analysis. In her research on domestic material culture in postwar Canada, for example, Joy Parr (1999) analyses a wide range of archival sources as well as interviews with industrial designers, industry experts and home-makers who 'furnished and equipped their houses between 1946 and 1968' (273). She interviewed women in British Columbia who contacted her after two newspaper articles described her research and made a request for interviewees. Together with Margaret-Anne Knowles, Parr 'began the interviews by collecting date of marriage, dates of birth of children, work histories of wife and husband, and making lists of their places of residence after marriage, with general descriptions of this accommodation, in chronological order' (273). The interviews then focused in detail on objects and furnishings within the home:

> We ... asked each person to describe the appliances and furniture she had acquired once married, in order of acquisition, noting, when they could recall, whether these goods had come to them as purchases, in trade, or as gifts. We tried to learn whether purchased goods had been bought for cash or credit, where, who had participated in what parts of the buying decision, and what factors had informed the choice. We asked for descriptions of these objects as they came into the home, and as they later were altered, and for judgments about how these things worked as tools and as parts of the household furnishings. We asked people to reflect on their priorities, as they changed over the years, both in terms of what they acquired and what they most valued in what they acquired, and to note conspicuous failures and successes.
>
> (273–4; for other research that draws on interviews about home furnishings, see Leslie and Reimer 2003, and Reimer and Leslie 2004)

Alongside interviews, ethnographic research on domestic material cultures also involves participant observation, as shown by Alison Clarke's research on the aesthetics of social aspiration for residents in a 'cosmopolitan but manifestly ordinary' street in north London (Clarke 2001b), and by Inge Maria Daniels' ethnographic research with two wealthy households in Japan (Daniels 2001). As Daniels explains, 'the material culture of the home is expressive of the changing social relationships of its inhabitants [and illustrates] the complexities, conflicts and compromises involved in creating a home' (205). Drawing on detailed descriptions of domestic material culture within the two case studies, and the ways in which objects are bound up with family identity, Daniels argues that 'the discourse surrounding the Japanese dwelling draws on an aesthetic and social ideal of order associated with the elite in feudal Japan. ... [but] real, lived-in Japanese homes are complex and cluttered' (225).

DATA SETS AND HOUSEHOLD SURVEYS

In this final section, we focus on the ways in which a range of quantitative sources and methods can be employed to study the home, concentrating on large, secondary data sets and household surveys. Data sets such as the census and other national surveys contain a wealth of information about housing, household demography and household finances, and their temporal, spatial and social differentiation. As we argue in Chapter 3, house and home are intimately connected and yet distinct. Although the imaginative, emotional and affective spaces of home cannot be quantified, data sets provide invaluable information for contextual analysis and to verify other findings for research on house and home.

Listokin et al. (2003) not only provide a very helpful review of the main American data sets that can be used in housing research (see Table 2.1), but also analyse the comparability of data across different sources. All of the data sets are sponsored or conducted by public institutions, and most can be accessed online.

Table 2.1   US data sets in housing research

Broad surveys of housing and demography:

Decennial Census

Summary files

Public-use microdata samples

Current Population Survey

American Housing Survey

Specialized data sets:

Survey of Income and Program Participation

Consumer Expenditure Survey

Panel Study of Income Dynamics

Survey of Consumer Finances

National Survey of Families and Households

Source: Listokin *et al.* (2003)

Like other sources, statistical data need to be interrogated, both in terms of the overall purpose of different data sets and the ways in which data are gathered, categorized, presented and analysed. As Listokin *et al.* explain, the differences across data sets 'represent the net effect of innumerable factors, including differences in respondent selection bias, panel and cross-sectional design considerations, specific interview protocols, question wording, definitions of household, and sample weighting procedures' (231).

Large secondary data sets such as the census have been particularly important in the study of household demographic change and household structure. Buzar *et al.* (2005) demonstrate the importance of the household scale in understanding demographic change in the developed world. Most of the processes associated with the second demographic transition – including 'the widening gap between total population and household numbers, accompanied by declining household sizes, falling fertility rates, and a wider palette of family and domestic situations' – 'emanate from the household scale, which has acted as a vessel for the increasingly fluid networks of kin and friendship' (414; also see Ogden and Hall 2004; Ogden and Schnoebelen 2005). As Buzar *et al.* argue, it is not only important to distinguish between 'family' and 'household' (also see Varley 2002), but also between 'home' and 'household'. As they explain, 'kinship and coresidence in the domestic group are not necessarily mutually interchangeable' and 'a single dwelling may contain several households, while in others one household may divide its time between different homes in different locations' (416).

The complexities of analysing household demographic change and household structure are also clearly evident in a wide range of research on development. The household is an important site for research and policy, particularly in terms of women-headed households (Dwyer and Bruce 1988; Varley 1996; Chant 1997). As Ann Varley explains, 'Household functions include co-residence, economic co-operation, reproductive activities such as food preparation and consumption, and socialization of children' (2002: 330). But, as she continues, households – like homes – are cultural constructs that are often closely bound up with, but are not the same as, 'the family'. Like the imaginative and material geographies of home that we explore throughout this book, households are not bounded units but, rather, their 'members' survival and well-being is also influenced by their connections with other households – kin, friends or neighbours' (2002: 330). Varley shows that 'Statistically, problems of definition and data quality are daunting' (331), and challenges the frequently repeated claim that one-third of the world's households are headed by a woman (also see Varley 1996). Summarizing data from the United Nations Women's Indicators and Statistics Database (1999) – which itself draws on the UN Demographic Statistics database, national census and survey data – Varley estimates that 'Excluding the developed countries, the figure is just under one in six households headed by a woman, and one in five overall' (331). Alongside such statistical analysis, researchers also employ qualitative research methods such as interviews, focus groups and participatory research to study women-headed households (see, for example, Chant 1997, which includes case studies of lone-mother households in Costa Rica, Mexico and the Philippines).

As well as using large, secondary data sets to analyse household demographic change and household structure, the household itself is also an important site for quantitative surveys and qualitative interviews, which can provide a wide range of material for studying the home. The British Household Panel Survey (BHPS), for example, was established in 1990 and includes 10,000 members, randomly selected across the UK. Panel members are interviewed each year about their work, income, health, attitudes, household living arrangements, housing and consumption (Buck *et al.* 1994; Berthoud and Gershuny 2000). BHPS is thus a large-scale, longitudinal survey that generates micro-data about social change. Through its annual questionnaire, the BHPS not only permits research on change at the individual or household level, but also enables the analysis of individuals over time (Buck *et al.* 1994: 4). In addition to national surveys such as the BHPS, many other researchers combine quantitative and qualitative research methods on a household scale. In their research on local labour markets and the gendered relationships between home and work in Worcester, Massachusetts, for example, Susan Hanson and Geraldine Pratt (1995) conducted semi-structured questionnaires with men and women in about 650 households in the Worcester area. As they explain, 'we randomly selected census blocks from subareas of the metropolitan area and then randomly selected five households per block to interview' (2003: 110). They were able to generalize from their quantitative analysis of coded answers '[b]ecause our sample was large and because sample households had been randomly selected in such a way that they were representative of the Worcester-area working-age population' (112). Moreover, Hanson and Pratt supplemented their quantitative research with qualitative analysis of open-ended interview data – the stories that respondents told about their experiences of local labour markets. Whilst the stories that people told about their experiences of finding work generated rich qualitative material, the quantitative analysis of household questionnaires enabled Hanson and Pratt to verify their findings and generalize them to the wider population.

CONCLUSIONS

The home is a rich site and subject for research. Geographers and other researchers seek to make sense of emotional and affective home-spaces alongside material cultures of home and the everyday practice of domestic life, and to explore the interplay of personal with shared and collective memories and experiences of home. This chapter has introduced some of the wide variety of methods and sources that researchers have used to study the material and imaginative geographies of home in the past and the present. Crucially, we have explored the ways in which material and imaginative geographies of home are closely bound together, as shown, for example, by the meanings and values associated with domestic material cultures, and the importance of textual and visual sources in understanding everyday, domestic lives. We have largely focused on qualitative sources and methods because we are concerned with the emotional resonance as well as the materialities of home, but we have also introduced quantitative research on home in relation to housing and households. We investigate the relationships between home, housing and households in more detail in the next chapter.

Studying the home raises important questions for research methodology more widely. Think, for example, about the location of research, and the particular challenges of conducting fieldwork not only close to home, but also within the domestic sphere. Rather than view the 'field' as discrete and distant, the home and everyday, domestic life are important sites for research, and places that might initially appear homelike are often unfamiliar. It is also important, but often difficult, to study the material and imaginative spaces of home that remain private, hidden or undocumented.

We draw on the wide range of sources and methods that we have introduced in this chapter throughout the rest of the book. In Chapter 3, for example, we consider domestic architecture, interviews and the material cultures of house-as-home, as well as quantitative research on housing markets and the monetary value of domestic labour. In Chapter 4, we begin by discussing two images of home, nation and empire, before drawing on a wide range of historical and textual sources in the rest of the chapter. We also discuss the recent film *Rabbit-Proof Fence*. Finally, in Chapter 5, we consider transnational geographies of home in relation to interviews, life stories, fiction, and material cultures of home.

Research box 1   HOME IN CONTEMPORARY INDIAN NOVELS
Alex Barley

My research focuses on the meanings and representations of 'home' in contemporary Indian novels. Choosing a literature-based thesis was deliberate: I wanted to explore how these novels dealt with the theme of 'home' within the wider social, cultural and political context of India.

To begin my research I read a wide range of contemporary Indian novels written in English to get a feel for different themes of home. In my readings, I found that many Indian novels were inward-looking, and both male and female authors were focusing on the home as a private space both in terms of a physical space where home is lived, and also as a symbolic space of emotions such as security, nostalgia, sacrifice, and a metaphor for the family and nation. Moreover, using 'home' as a major research theme provided me with an overarching concept through which to explore these Indian novels, because 'home' is a concept that is intertwined with other issues such as colonialism, nationalism, gender, class and caste. Some of the research questions I developed to explore these issues further include:

- What are the themes of home in the novels?
- How have colonial and nationalist discourses in India shaped ideas of home?
- How do Indian novels conform to or subvert these discourses?
- How are these discourses manifested at different scales including local, national and transnational levels?

Rather than survey the field of Indian literature, I decided to answer these questions by focusing on specific novels in order to undertake a more in-depth analysis of the texts. The novels and authors I selected were: *Freedom Song* by Amit Chaudhuri, *Sultry Days* by Shobha Dé, *Fasting Feasting* and *Fire on the Mountain* both by Anita Desai, *Tara Lane* by Shama Futehally, *The Blue Bedspread* by Raj Kamal Jha, *The Romantics* by Pankaj Mishra, *Ancient Promises* by Jaishree Misra, and *A Fine Balance* by Rohinton Mistry. I chose these partly out of personal preference – I enjoyed the novels – but also because they raised interesting issues about home that I wanted to explore further; some of these issues are:

- a dissatisfaction or frustration with home – a desire to escape or run away from home;
- nostalgia for a past home – this might be about a specific time period such as childhood, or a location such as a village;
- a rupture – displacement from home caused by an event such as Partition or migration; and
- the rhythms of daily life in the home – family life, growing up, marriage.

What is the wider significance of these themes? Well, the first of these themes – dissatisfaction with home – is a common theme across the novels and takes many forms. In the case of the novel *Fire on the Mountain*, the elderly widowed protagonist Nanda Kaul is an angry woman full of resentment about her family and married life. My interpretation of her resentment includes considering the personal circumstances of Nanda Kaul, but also relates it to the broader issues regarding conflicting imaginings of the self and the nation about women, the home and family life. So, using the theme of home, I show how these Indian novels produce versions of India that stretch 'home' from something that is of personal or family concern into wider national significance.

Alex Barley is currently temporary lecturer in Geography at the University of Durham, where she recently completed her PhD. Her thesis is entitled 'At Home in India: Geographies of Home in Contemporary Indian Novels'.

Research box 2   MEMORY-WORK IN THE MARGINS
Caitlin DeSilvey

The ramshackle sheds and cellars of a homestead in the foothills of the Rocky Mountains, crammed with the remnants of a century's habitation, presented the unlikely setting for my PhD research into material cultures of an abandoned home. The homestead's clutter – a miscellany of commonplace household and farm objects, stirred into a surreal stew – challenged me to come up with methods to explore it without clearing away the unpredictable effects that drew my attention to it in the first place. I framed my research as a salvage of cultural memory, rather than a narration of a particular history. The work I carried out could be described as a form of domestic archaeology (though in this domestic realm the alternate domesticities of rodents and microbes had long ago scrambled any ordering principle). Despite the apparent disorder, the contents of the structures retained traces of the complex syntax of home, an interlocking structure of latent meaning in which objects provide material and symbolic links to subjective states of being and doing (Miller 2001). My methodological approach involved an imaginative reconstruction of the memories materialized in the homestead's discarded belongings.

If the material world is intimately bound up in the production of self – and of home – this understanding takes on a peculiar cast when research works with the things people leave behind (Buchli and Lucas 2001). Objects not deemed worthy of saving carry with them a residue of experience unmediated by practices of deliberate remembrance and display. The things I found indexed, for example,

Figure 2.6   Contents of the homestead junk drawer. Reproduced courtesy of Caitlin DeSilvey.

endless hours of spent labour. Junk drawers clogged with scavenged odds and ends recalled cycles of mending and making-do (Figure 2.6). Saved string, wound into dozens of neat balls, evoked a centripetal force to the core of the household. An orientation to the world outside the bounds of the farm appeared in mouldy delivery records, in jars of mineral soil collected from distant hillsides to feed a speculative hunger, in glossy images on the pages of mouse-nibbled National Geographic magazines. Some of the objects evoked the ghosts of homes lost, or never attained. The iconography of a rural idyll – a pail of milk in a daisy-strewn field, a rosy-cheeked child hunting rabbits – appeared in clippings on the dusty walls. Other materials exposed the home's uncertain foundations. The root cellar turned up countless documents of money owed and money due, mortgages and meagre bank statements. Grandchildren who lived on the homestead during the Depression left behind a scattering of marbles, and a drawing of the homestead's kitchen (Figure 2.7), the centre of a temporary home bathed in thick beams of yellow-crayon lamplight.

Figure 2.7 A child's drawing of the homestead kitchen. Reproduced courtesy of Caitlin DeSilvey.

An 'irreducible scrappiness' characterized the memories I was able to salvage from the homestead's debris (Smith 1988). There was no whole story to tell, only many small stories. The objects and their stories rendered 'home' as a complex, and sometimes contradictory, structure of affection and aversion, rootedness and restlessness. Marginal things, cut loose from the careful maintenance of memory that takes place in photograph albums and on keepsake shelves, remembered their past otherwise.

Caitlin DeSilvey is a research fellow in the Open University Geography Discipline. Her doctoral thesis, 'Salvage Rites: Making Memory on a Montana Homestead', developed a set of alternative methods for the interpretation of residual material cultures. She also writes and works on the history of allotment and community gardens, the geographies of contemporary art practice, and the heritage management of abandoned settlements in the American west.

3

RESIDENCE: HOUSE-AS-HOME

House and home are often conflated: home is the place, dwelling, or shelter in which people live. In Chapter 1 we introduced two criticisms of this conflation of house and home. First, we pointed out that home does not have to be attached to a house; imaginaries of home can be connected to numerous places at multiple geographical scales. Second, we argued that since the connections between house and home are *made*, analysis needs to demonstrate those connections rather than assume them. *How* and *why* do built forms – houses or dwellings – become identified as home? What are their social and spatial characteristics?

The processes through which houses become home are the focus of this chapter. We do not aim to provide an exhaustive overview of all dwellings and all imaginaries of home. Instead, we provide a critical geographical analysis of house-as-home. We begin with a section on home economies, in which we trace the multiple connections between housing, economic processes and economic relations in order to highlight the economic importance of home and the inter-relations between the spheres of home and work. In the second section we investigate how a normative notion, or ideal, of home is materialized in the form of dwelling structures, and we elaborate the socio-spatial characteristics of 'ideal' or 'homely' homes. The chapter finally provides a discussion of houses and experiences of dwelling considered to be 'unhomely'. Running throughout the chapter are the key themes of a critical geography of home. In addition to our focus on the links between material and imaginative homes is an interest in the identity politics of home. Whereas the spatial layout and perception of many forms of dwelling correspond to dominant ideologies of home, these ideas are constantly resisted and recast, often through home-making practices. Thus we use people's house-based practices to illustrate their active engagement with, and recasting of, the social characteristics of home. Finally, we are concerned with the scale politics of home. Certainly much of the discussion of this chapter confines itself to houses and domestic practices, but we also show the multiple scales of these places and practices, whereby home can be stretched beyond houses or reduced to the body.

HOME ECONOMIES

Just as home is not divorced from economic relations, houses are the source and location of diverse economic relations. For example, the provision of housing is monetized in many societies, and parts of non-monetary systems of exchange in many others. In the main, houses are commodities, with the implication that houses become assets and sources of wealth for some, or associated with poverty for others. We consider three different aspects of economic meanings and connections of home in this section: the monetary significance of housing to both national economies and individuals (and hence the imbrication of the domestic with other scales); the intertwined social, economic and cultural distinctions between people or groups of people that emanate from home economies; and home-based work.

The economic significance of housing

In most societies people have to pay for a physical structure that they can call home. Consequently, the provision of housing is connected to wider circuits of capital accumulation and labour processes. A substantial amount of money in many nations is spent on building and renovating houses. In the United States, nearly two million new dwellings are built each year (US Census Bureau 2004), almost one million in Japan (Web Japan 2004) and around 200,000 in each of Australia, the United Kingdom and Canada (HIA 2004, ODPM 2004, Statistics Canada 2004). Housebuilding also provides employment for many people. According to a number of measures, economic activities associated with the building of new houses are growing. The proportion of national income derived from housing-related activity, for example, in Australia, Britain and the

United States has been increasing over time. Another measure of the economic significance of housing is the percentage of private wealth held in housing. In Australia, residential property accounts for almost 60 per cent of total wealth held by individuals; the comparable figure for the United States is 20 per cent (Badcock and Beer 2000: 97). Japan is the nation with by far the most private investment in housing, with 70 per cent of individual wealth accounted for by residences (Badcock and Beer 2000: 97). Though variable across different countries, it is nonetheless the case that domestic building remains important to national economies and the wealth of individuals.

One further aspect of the economic significance of housing is captured in Figure 3.1. Though an exaggeration, this figure aptly draws our attention to the economic importance of house-as-home as a site of consumption. In making houses homes, in carrying out domestic activities, in nurturing and caring for family members, goods and services are bought (see also Box 3.1). Such consumption dramatically increases the amount of expenditure related to dwellings, and hence multiplies the economic significance of home. Indeed, some people have argued that since the middle of the twentieth century house-related expenditure has been critical to the very survival of capitalism. In his analysis of housing in post-war Australia, for instance, Alastair Greig (1995) captions Figure 3.1 'filling the Fordist container'. By this he means that homes are containers for all the mass-produced items of Fordism such as refrigerators, televisions and lawnmowers. Home-related consumption remains important in the twenty-first century, where house-related expenditure continues to rise, though this time through 'home theatre systems', personal computers and the like.

Housing tenure, social divisions and identities

There are three main ways of paying for housing: purchasing or owning (termed owner occupation), renting from a private landlord (private rental), or renting from the state (public rental). Tenure mixes vary considerably across developed countries, as shown in Table 3.1. Cutting across these national differences are the social divisions created and sustained through housing tenure. In other words, how a dwelling is paid for is related to, and influences, identities of class, gender, race and other social divisions. Owner-occupiers (who either own their dwelling or are purchasing it), are more likely to be employed, white, middle-class, middle-aged and educated. Private renters, especially long-term renters, are more likely to be

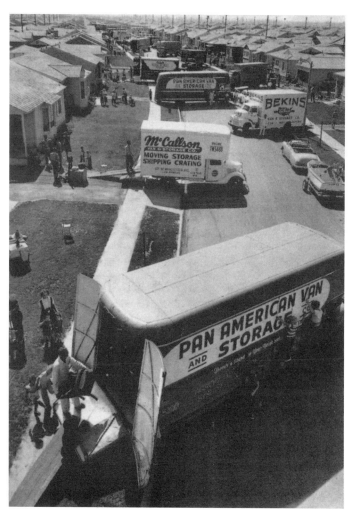

Figure 3.1 Multiplying the economic significance of housing. Expansive view of newly built houses jammed side by side, 1952. Photograph by J. R. Eyerman, Time Life Pictures, Getty Images. Reproduced by permission of Time Life Pictures / Getty Images.

unemployed, young and in non-family households. For those renting from the state, social disadvantage is even more pronounced (Badcock and Beer 2000). Tenure divisions may also exacerbate levels of social inequality through the additional social and economic benefits that flow to owner-occupiers (Yates 2002: 587). Many countries have favourable tax treatment of some aspects of ownership, and owners are able to derive capital gains from their houses and hence further improve their financial position. Private and public renters, on the other hand, do not have security of tenure, are not able to reap economic gain from their dwelling and may be associated with poverty after retirement (Yates 2002: 588).

Race is another dimension that is related in complex ways to housing tenure. With respect to housing options and conditions, black and minority ethnic communities in Britain 'on balance fare disproportionately badly when compared with the majority "white" population' (Ratcliffe 1998: 807), being less likely to be homeowners or have security of tenure. Similarly, in the United States access to home ownership is racialized through practices of mortgage lending. Discrimination in processes that make mortgage credit available limit black and minority groups' potential to become owner-occupiers (see Holloway and Wyly 2001; Newman and Wyly 2004). As with all processes of inclusion and exclusion, the links between housing tenure and social disadvantage are complex and remain important.

Squatting is another form of tenure with associated social divisions. Squatting is the practice of living in and perhaps modifying a dwelling without the consent of the owner. It is widespread in a diverse range of large cities such as Hong Kong, Amsterdam, Mumbai, Rio de Janeiro and Nairobi (see Neuwirth 2004; Smart 2002). By definition illegal, squatting primarily exists outside the formal economies discussed so far in this section. Squatters, according to Neuwirth (2004: 10):

Table 3.1   Housing tenure in different countries

Country	Owner-occupation (%)	Private rental (%)	Public rental (%)
Australia	70	24	6
Britain	68	11	21
United States	65	30	5
Canada	63	30	7
Denmark	60	21	19
France	57	19	24
Germany	39	42	19
Sweden	41	21	38

Source: Badcock and Beer (2000): 2

have created a huge hidden economy – an unofficial system of squatter landlords and squatter tenants, squatter merchants and squatter consumers, squatter builders and squatter labourers, squatter brokers and squatter investors, squatter teachers and squatter schoolkids, squatter beggars and squatter billionaires.

Tenure divisions between owning and renting also have significant cultural purchase, especially in terms of ownership. Ownership is most closely identified with dominant notions of home. Notably, ownership is termed 'home ownership' rather than 'house ownership', signalling that ownership is synonymous with home. Those buying a house are presumed to be properly capable of making home, of creating a place that is secure, comfortable and welcoming. It is ownership that makes 'real and possible the control, the security, the status, the family life that people seek through their houses' (Rakoff 1977: 94). It is also ownership that becomes symbolically connected with some national identities, a national identity that is dependent on a particular definition of home (as we outline in Chapter 4).

Tenure divisions and associated identities are both reproduced and contested through home-making practices. The accumulation of wealth through home ownership can be the outcome of gendered home-making practices. Housing researchers have identified a 'housing ladder' in which smaller, cheaper houses in not-so-good locations are to be found on the lowest rungs and larger, more expensive houses in better locations occupy the higher rungs. Many individuals tend to climb this ladder during their housing careers (see Box 2.2 on housing careers). Dowling's (1998a) study of Vancouver homeowners showed that climbing this housing ladder was dependent on the work of women in these households. The decorating and renovation work done by the women ensured a substantial capital gain in one house. The women then initiated and organized moves to 'better' locations. In extreme cases the women became owner-builders or project managers overseeing the building of the new house in order to maximize their financial gain. This example demonstrates how home-making practices underpin some of the dominant ideologies of home, in this case the economic benefits of home ownership. But the example also demonstrates the reworking of these ideologies. Whilst women are associated with home, it is not just through family. In this example, women's work as builders and financial managers were activities more usually associated with men, and hence broaden gendered associations of home.

Home-making practices may also contest dominant meanings of home ownership. Self- or owner-building has been used by low-income people to bypass the capitalist provision of housing. Building their own home was key to many working-class people becoming home-owners in early twentieth-century Toronto, for example (Harris 1996). Similarly in Sydney in the 1950s, working-class families were dependent on their own labour, as well as the labour of their friends and families, to attain home ownership. As one man recalled:

> We used ammunition boxes to form the concrete pillars and had to buy these small weatherboards for outside as fibro was impossible to obtain. Once the basic home was up, we built the garage. I would make the frames myself with whatever timber I could get but needed my father and Lois [his wife] to help stand it up. As the children grew bigger we needed additional rooms. By the early 1950s, building materials were easier to get and we could build the remainder of our house ... Lois was my labourer on the site. She creosoted the timbers, lugged buckets of water and mixed the cement for our pathways.
>
> (Allport 1987: 104)

Non-profit agencies have long been involved in assisting low-income people into home ownership. A well-known recent and widespread example is Habitat for Humanity (see www.habitat.org). Since 1976 the organization has built more than 175,000 houses around the world using volunteer labour and donated money and materials. Families in need of shelter apply to the organization for housing. If accepted, they pay a deposit and agree to make monthly mortgage payments. They also have to invest their own labour into the building of their own house and the houses of others. Like self-building, Habitat for Humanity facilitates home ownership through the labour of owners. Also like self-building, it is a shared labour in that owners help in the building of other houses as well as their own.

Home as work

House-as-home is also a space of production or work. When household tasks such as cooking, cleaning, and caring for children occur outside the home they are classified as work and remunerated. Yet household or domestic work that occurs in the home as part of relationships of kin and caring is, in the main, not remunerated (but see pp. 97–9) and not counted in the compilation of national accounts (Luxton 1997). Attempts to measure time spent in domestic work, and ways of estimating its economic value, are varied and contested. All measures, nonetheless, show that home-based unpaid work is substantial both economically and in terms of time. For example, in twelve countries in the Organization for Economic Cooperation and Development (OECD), adults spend more time in household work (26 hours per week) than they do in paid work (24 hours per week) (Ironmonger 1996: 45). Attempts to value this unpaid work in economic terms variously suggest that the value added by household production was almost as much as the value added by market production (Ironmonger 1996: 52), or that 60 per cent of personal disposable income would be needed to pay for this work (Luxton 1997: 437).

Gender is a critical differentiating factor in home as a workplace because most domestic work is carried out by women. In the Netherlands, for example, women spend 35 hours per week on household tasks, compared to 20 for men (Ironmonger 1996: 52). In Britain, fathers spend on average 23 hours per week cooking, cleaning, washing, shopping and caring for and driving children, whilst mothers spend 62 hours on the same tasks (Chapman 2004:105). These gender differences in home-based work become the basis for further inequalities, limiting, for example, women's participation in paid work and education. One response has been to argue for the remuneration of domestic labour. The 1960s and 1970s saw domestic labour expressly politicized through campaigns for wages for housework (see McDowell 1999), whereas others maintain that domestic technology reduces the burden of domestic work (see Box 3.1, and Research Box 3 by Helen Watkins).

Box 3.1 DOMESTIC TECHNOLOGIES

Technological devices and advances have long been used in the work of maintaining and reproducing home. In the late nineteenth century indoor plumbing, gas and then electric lighting, and central heating made the attainment of a comfortable home less physically demanding (see Rybczynski 1988; also see Box 3.3 and Research Box 3 by Helen Watkins). The use of domestic technologies for the work of home became even more pronounced in the twentieth century as the availability and use of appliances became more wide-spread. Washing machines, refrigerators, vacuum cleaners and dishwashers were just a few of the technologies that transformed cleaning the home, microwaves and food processors altered cooking, and gardening became less physically demanding with the use of power tools and lawnmowers (see Chapman 2004: 101–2). Both manufacturers and government often advertise these devices as 'labour-saving', though the evidence of whether this is actually the case is sparse. For Ruth Schwartz-Cowan (1989), the main impact of domestic technologies has been an increase in the work of home, especially for women. Whilst these technologies have made cleaning and cooking processes simpler and less physically arduous, they have also increased expectations of cleanliness. Thus clothes are now washed more frequently and microwaves have increased the demand for flexible and instant meals (Chapman 2004: 103). As Tony Chapman summarizes, 'the introduction of household tech-nology has been both a blessing and a curse in terms of its time-saving advantages' (104).

The application of scientific rationalities to home life is another form of domestic technology. Rather than appliances, these are tech-nologies of behaviour and morality that transform home-based work and home-making practices. Home in early twentieth-century Australia, for example, was characterized by attempts to 'rationalise' the domestic, 'to extend the principles of science and instrumental reason to the operation of the household and to the management of personal relationships' (Reiger 1985: 3). Scientific rationality was applied to many different spheres of home life:

> Efforts to introduce technology to the household and to define the housewife as a 'modern', 'efficient' houseworker; to change patterns of reproduction by placing contraception, pregnancy and childbirth under conscious, usually professional, control; to alter childrearing practices in the light of 'hygiene', seen as both physical and mental; and to bring sexuality out from under the veil of prudery and silence.
>
> (Reiger 1985: 2)

A final important domestic technology is television. Television is a medium that not only portrays home life (and see Box 2.7 on home and film), but also one that is watched at home and incorporated into domestic life. It sometimes occupies a central place in our house, and may be the central home-based activity for many (see our discussion of older people and home on p. 113). Television use is subject to home rules and routines as well as negotiations between parents and children (Silverstone 1994: 38). Television use is also gendered. David Morley found that men dominate television use at home, controlling 'programme choice, viewing style, the planning or unplanning of viewing, the amounts of viewing, television related talk' (Morley 1986, as summarized in Silverstone 1994: 39). Moreover, television is an ideal example of the porosity of home. Television literally brings the 'public' world into the home:

> television itself, as medium and as message, will extend and plausibly transform a household's reach, bringing news of the world of affairs beyond the front door; providing narratives and images for identification, reassurance or frustration; affecting or reinforcing the household's links with neighbourhood and community; and locking the household ever more firmly into an increasingly privatized and commodified domestic world.
>
> (Silverstone 1994: 50)

Today other domestic technologies such as the home computer and internet are fulfilling similar roles (see Holloway and Valentine 2001, and Box 5.3 for further discussion).

Not all domestic labour is unpaid. The late-nineteenth-century middle class in Britain and elsewhere, for example, functioned through the paid labour of domestic servants, as we shall see in the next section. After fifty years in which the number of households employing domestic servants fell, some research is now suggesting that paid domestic work (or outsourcing of domestic work) is again on the rise in middle-class households. A recent estimate, for example, is that there are 60,000 au pairs in Britain (Cox and Narula 2003; Gregson and Lowe 1994). Certain forms of household work are more likely to be outsourced than others. In a comprehensive Australian analysis, Bittman *et al.* (1999) find that some outsourcing of domestic food preparation (in other words, eating out at restaurants or buying take-aways) has become almost universal. Similarly, household expenditure on childcare is increasing. Paid childcare work often occurs inside the home. Research conducted in Britain by Louise Gregson and Michelle Lowe shows that most middle-class parents prefer to employ a nanny rather than a childminder as the care would then be located within the parental home rather than in someone else's home (1994). Like other caregiving work such as nursing and home-care for older people, childminding is usually poorly paid. A US government survey conducted in 1998 found that the average wages for early education and care staff ranked 757 out of 774 occupations surveyed. As summarized by Geraldine Pratt, 'Americans pay more to those who attend their parked cars than to those who attend their children' (2003: 581). Only a small proportion of households, however, pay others for cleaning, laundry or gardening. The use of paid domestic labour may be increasing, but it remains far from widespread.

Because of its location in a 'home' – predominantly understood as a space of private, familial, non-economic relations – paid domestic work is a somewhat unusual form of work (Cox and Narula 2003). In particular, relations between employers (the householder, typically a woman) and employees are crosscut by imaginaries or ideologies of home. As a result, relations between domestic worker and householder become 'quasi-familial' or akin to, but different from, familial relations. A 'friend-like', negotiated relation may be constructed, with attempts to make the worker treat the house as their 'home' and use the space freely. According to one employer,

> If you're living with an au pair then, for me anyway, it's crucial that person becomes part of the whole family. And, therefore, the first word is 'anything that is in here' – and this is quite genuine – 'is yours'. So it could be the fridge, the TV or whatever.
>
> (quoted in Cox and Narula 2003: 339)

Much more common, according to Cox and Narula (2003), is for the householder to become parent-like and treat the worker like a child. Strict rules about use of space and time within the house ensue. These include rules about when and where guests are welcome, and eating with the children rather than the adults.

These forms of paid domestic work are fascinating instances of the diverse ways in which domestic relations are stretched beyond the home as well as the ways in which homes are constituted through their relations with the 'public'. The home-based activities of paid domestic workers such as nannies extend the processes of reproducing family life beyond immediate family members and their home-spaces. In some cases, as we discuss in detail in Chapter 5, reproduction is also stretched transnationally. These activities also bring non-kin relations into the realm of home, principally labour relations but also relations with the state. Box 3.2 discusses some of the implications of previously state-based caring practices for notions of home. Paid domestic work is regulated (and sometimes encouraged, as we discuss in Chapter 5) by state agencies. Other forms of home-based employment similarly stretch home and connect spheres of home and work. These include home-based garment production, which rose to prominence at the beginning of the twentieth century, and which remains prevalent at the beginning of the twenty-first (see Pearson 2004 for an international overview). Teleworking – where an office worker works at home and is connected to the office through telecommunications – is a more recent variant (see Ahrentzen 1997). In both cases, home is a site of both production and social reproduction.

Box 3.2   CARING WORK AT HOME

Homes are sites of caregiving, principally through kin. As state financing of health and welfare services has declined in the west over the past decades, non-kin caring increasingly occurs at home. People with illnesses are cared for by medical professionals in their home environments, people with learning disabilities or mental illnesses live in family homes as well as institutions, and terminally ill people die at home rather than in a hospice (see Dyck *et al.* 2005). Both government policies and changing attitudes towards illness, dying and disability have contributed to this shift towards 'community care', and pose a number of challenges to the dominant meanings of home, especially its perceived 'privateness' (see Dyck *et al.* 2005).

   The home-based provision of care for the terminally ill provides one insightful example of the simultaneous unhomeliness and homeliness of home. In the United States, since the 1980s, there has been a rise in the number of people dying at home, linked to changes in the funding of care for the dying and the commodification and rationalization of health care (Brown and Colton 2001). Home as a site of terminal caregiving is a paradoxical space. Michael Brown's (2003) research on hospice care in Seattle shows the material and imaginary paradoxes that emerge when a traditionally unhomely practice – dying – occurs there. As a hospice, home becomes coded as both good and bad – a familiar, non-institutional setting, but also a site of unresolved family tensions. When home becomes a place of terminal caregiving it remains associated with comfort and control, but there is also lack of control, as strangers (publicly provided caregivers) and/or relatives may move in. And terminal caregiving at home sees the co-presence of professionalized and familial practices of caregiving.

For Isabel Dyck and her colleagues (2005), home in the context of long-term health care is similarly paradoxical, 'simultaneously both and neither private, public, individual or social' (181). Home becomes much more public, a space entered and sometimes controlled by health professionals, as they care for the residents' needs every day. Home-making activities nonetheless persist. Dyck recalls the attempts of one woman to personalize her bedroom. She made it home-like and individual by matching clothes and bedlinen, and positioning her bed so that she was able to see who was entering. These spaces are both homely and unhomely at the same time, riven with feelings of belonging and attachment, alienation and detachment.

'HOMELY' HOMES

A central feature of imaginaries of home is their idealization: certain dwelling structures and social relations are imagined to be 'better', more socially appropriate and an ideal to be aspired to. It is these dwelling structures and social relations that become 'homely homes'. Public discourse – in the media, in popular culture, in public policy – presents a dominant or ideal version of house-as-home, which typically portrays belonging and intimacy amongst members of a heterosexual nuclear family, living in a detached, owner-occupied dwelling, in a suburban location. In this section we outline the characteristics of these ideal or homely homes. We begin by sketching an answer to the questions of where ideal homes are presumed to be located and the built form they are presumed to take. We then deploy our critical geography of home to provide a sceptical and more complex understanding of these 'homely' homes, with our discussion focusing on the home-making practices that reproduce, contest and materially transform them.

Locating homely homes

Ideal homes are culturally and historically specific (see Oliver 1987 and Box 3.3). Within the frames of contemporary western domesticity, an ideal home is considered to be a suburban one. A detached or semi-detached house, situated on a large block of land on a city's outskirts, built in the image of a family and individually owned, is considered the dwelling that can most easily and 'naturally' become home. The construction of the suburban home as ideal can be traced to the growing separation of public and private spheres in late-Victorian Britain. For much of the nineteenth century most urban dwellers established homes in central city locations, close to workplaces, and often in terraced houses. During the nineteenth century these neighbourhoods and the attached-house form became reviled by the middle class, in part because of their lack of separation of public and private spheres (Davidoff and Hall 2002). A new house form, in a new location, in which historically specific social relations were enacted, thus emerged, facilitated by the development of suburban railways. The location – in open countryside beyond the confines of the industrial city – was key to its social meaning and home-making practices. Homes in the countryside signified the positive valuation of proximity to nature, and, in particular, a belief that nature was most beneficial to the raising of children. Location away from city workplaces was also a manifestation of the socio-spatial separation of spheres of home and work. If home was to be a respite from work (though bear in mind the criticisms of this distinction we outlined in Chapter 1), then the suburban home was also constructed as spatially separate from work. Suburban house design also reflected this separate-spheres ideology and the ideals of bourgeois family life. It spatially demarcated public and private, masculine and feminine spaces (Madigan and Munro 1999a).

Box 3.3   HOME AS IMAGINED AND MATERIAL COMFORTS:
RE-READING WITOLD RYBCZYNSKI'S *HOME*

Witold Rybczynski's (1988) modestly titled *Home: A Short History of an Idea* is a book-length treatment of the material manifestation of home. This book was important in bringing home to the attention of scholars from many disciplines, though from a very different theoretical and empirical perspective from this book. He takes for granted, even romanticizes, connections between home, individuals and family:

> Domestic well-being is too important to be left to the experts; it is, as it has always been, the business of the family and the individual. We must rediscover for ourselves the mystery of comfort, for without it, our dwellings will indeed be machines instead of homes.
>
> (232)

Nonetheless, Rybczynski's book can be usefully interpreted as demonstrating two key elements of a critical geography of home.

First, he provides an example of the geographical and historical specificity of meanings of home. He traces the 'pre-history' of the modern home – from where, when and in what social conditions the idea of the modern home emerged. Home has a number of different elements: privacy, domesticity, intimacy and, most importantly for Rybczynski , comfort. Home as a space of privacy and domesticity emerged in the eighteenth-century home. As the British house began to accommodate fewer occupants, and was no longer a place of work, it became a place for personal, intimate behaviour. This notion of intimacy was reinforced by changing attitudes towards childhood in which children occupied houses, and were not treated as adults, for much longer periods of time (77). Domesticity, as concern for and pride in home environments, Rybczynski traces to the eighteenth-century Dutch house, where furniture was used to convey wealth and the home became the setting for the social unit of the family. Comfort was the last of the modern meanings of home to emerge, after privacy and domesticity. Whereas the medieval home didn't lack comfort, it was not explicit:

> Rooms hung with richly decorated tapestries were poorly heated, luxuriously dressed gentlemen and ladies sat on plain benches and stools, courtiers who might spend fifteen minutes in elaborate greeting slept three to a bed and were unmindful of personal intimacy.
>
> (34)

Comfort – which he states should be defined by inhabitants rather than architects, engineers or designers – is loosely translated as feelings of ease and relaxation. Comfortable furniture began to appear in houses, with 'cosy corners' established in drawing rooms (118). As sitting became a form of relaxation in France ('people sat together to listen to music, to have conversations, to play cards'), new 'comfortable' chairs that accommodated the postures of the time became more widely used (97). By the 1890s, the key components of the modern idea of home had emerged.

So too, by the 1890s, had the technological devices (central heating, indoor plumbing, electric light) to facilitate these modern ideals of home (219), which brings us to the second of Rybczynski's contributions to a critical geography of home: his elaboration of the inter-relations of ideas, technology and practice in the making of a comfortable modern home. In our framework, he documents the intersection of materialities of home, imaginaries of home, and home-making practices. Material transformations were important, especially technological innovations and the work of furniture designers and architects. Equally important were imaginaries of home, ideas of what home should be. And finally important were the ways rooms were used: 'the way that rooms looked made sense because they were a setting for a particular type of behaviour which in turn was conditioned by the way people thought' (219). For example, the mechanization of the home made comfort easier to obtain. Technological advances, principally electricity, provided the basis for improvements in comfort:

So many aspects of the modern home that we take for granted date from this period – the small size of the house, the correct height for work counters, the placement of major appliances to save needless steps, the organisation of storage.

(171)

Homes of the past are rather more palatable than contemporary homes for Rybczynski . Contemporary modernist interiors, with their 'conspicuous austerity', deny the connections between home materialities, imaginaries and practices:

It represents an attempt not so much to introduce a new style – that is the least of it – as to change social habits, and even to alter the underlying cultural meaning of domestic comfort. Its denial of bourgeois traditions has caused it to question, and reject, not only luxury but also ease, not only clutter but also intimacy. Its emphasis on space has caused it to ignore privacy, just as its interest in industrial-looking materials and objects has led it away from domesticity.

(214)

This passage is problematic in the ways that home-making practices and the ideas of comfort, privacy and domesticity are deemed more important than style. Nonetheless, Rybczynski's work remains important for its linking of material and imaginative geographies of home.

In the twentieth century, the suburban house as ideal home remained a product of ideologies of separate spheres and the prominence given to relations within the nuclear family as the cornerstone of social life (see Box 3.4). In the postwar period numerous layers of government policy contributed as well. In the United States, the mass production of suburban houses, and their idealization, was propelled by national housing policy (see Checkoway 1980). Similarly in Australia, the construction of suburban houses was encouraged as part of a policy to increase the national birth rate (Allport 1987).

Box 3.4   SUBURBAN HOUSE DESIGN OVER THE TWENTIETH
CENTURY

As Alison Ravetz (1995: 18) reminds us, 'the suburban home and the
suburb typify the twentieth century'. Though this period was one of
profound social and economic change across the world, the basic
form of the suburban home remains unchanged. It is still imagined
and constructed as a space for the nuclear family, a respite from work
and a signifier of an individual's status. Nonetheless, three aspects
have changed in subtle ways. First, the suburban home has expanded
in size, despite the decline in family size across the western world.
New houses in the United States have grown from 800 square feet
and one bathroom in the middle of the twentieth century, to 2,250
square feet and two and a half bathrooms at the beginning of the
twenty-first (Hayden 2002: 194). The pace of change has been most
dramatic from the 1980s onwards. In Australia in the mid-1980s
houses were, on average, 150 square metres, but had increased to
more than 200 square metres in 2004 (Australian Bureau of Statistics
2004). Second, the exteriors of suburban houses have changed with
alterations in fashions and tastes. For example, the original Levittown
house design was termed Cape Cod, with a name designed to connect
the house with the past (see Kelly 1993 and discussion on p. 117–8).
Indeed Levittown, which some claim to be the first mass-produced
suburb with mass-produced homes, is a wonderful case study of the
suburban   home   (see   www.uic.edu/~pbhales/Levittown.html).
Finally, as changes in family living have occurred, so too has suburban
house design. These are shown in the diagram, which is based on
twentieth-century house design in suburban Australia. In the middle
of the century the provision of separate public and private places
meant the separation of family activities and members. This slowly
gave way to the larger, open-plan design of the 1980s, with shared
communal spaces (Dovey 1992; Madigan and Munro 1999b; Attfield
2000) reflecting ideals of family togetherness. The 1990s onwards
has seen minor alternations to this schema. Whilst an open-plan
design remains favoured, some generational separation has reap-
peared in the form of 'parents' retreats'. These spaces, typically
located close to parents' bedrooms and away from 'family spaces' are

designated adult spaces in advertising. These house designs also reflect the priority given to providing spaces for families and children. Spaces for non-family members are not explicitly provided, nor are spaces for work or other extra-familial activities.

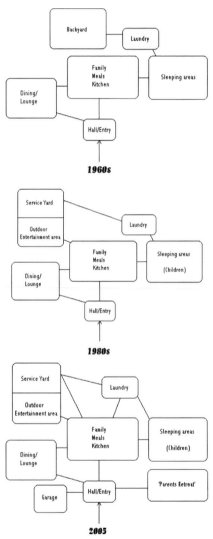

Figure 3.2   Spatial organization of Australian houses, 1960s to 2005. Adapted from Dovey (1992) and author's survey.

Paralleling the social construction of the suburban house as an ideal home is the presumption that a city dwelling cannot accommodate normative ideals of home. They are not for families, are sites for the enactment of non-familial gender relations, and are occupied by social groups other than the middle class. City homes also take a different spatial form. Rather than spatially separate, more than one dwelling occupies a parcel of land and individual dwellings share walls and entrances. The notion that apartments are unhomely is historically and geographically specific. In the United States and the UK this notion is linked not only to middle-class attempts to produce an ideal of home in their own image (see Marston 2004), but also to the negative associations of apartment and high-rise living (see Box 3.5).

Box 3.5 APARTMENTS, THE STATE AND THE SOCIAL CONSTRUCTION OF HOMELINESS AND UNHOMELINESS

For much of the twentieth century the high-rise apartment has been understood in terms of an absence of home. One of the reasons for this perception is that high-rises have commonly been built by governments and used to provide public housing. In mid-twentieth-century Britain, the United States and Australia, governments cleared inner-city neighbourhoods of working-class housing and constructed high-rises instead. The story of the high-rise in Melbourne, Australia, provides a useful illustration of connections between social and political conditions and the construction of an unhomely space. In the mid-twentieth century in Melbourne, a concerted effort was made to rid the city of inner-city working-class housing, which was characterized as slums. It was feared that these neighbourhoods bred disease and perpetuated people's low moral standards. High-rises were hence built to re-house working-class inhabitants, because they were able to efficiently (read cheaply) house large numbers of people (Costello 2005). High-rises, because they were modern, were also presumed to underpin higher and therefore more acceptable standards of morality and respectability. In intent, then, the high-rise was to become home. Yet it was not long before the 45 high-rise blocks constructed during this period were re-scripted as unhomely:

> High density living is said to produce family breakdown, delinquency and a variety of social problems ... provides a bad environment in which to raise children, and [creates a situation whereby] families are denied the privacy to which they are entitled. It is commonly suggested that these people are thrown into dangerous idleness when they have no garden to tend.
>
> (Stevenson *et al.* 1967: 8, cited in Costello 2005: 53)

Imaginings of home quickly became connected, in other words, to negative perceptions of the worth of inhabitants. These connections have also been made in the United States and the UK, where they are more strongly connected to racial stereotypes (Murray 1995).

Emphasized in these discussions of high-density public housing is alienation, whereby it is impossible to feel at home, or make home, in these dwellings. Theoretically, a critical geography of home draws our attention to the simultaneity of belonging and alienation. Some research, especially that which pays close attention to home-making practices, allows both elements to be recognized. Exemplary here is a piece of research conducted almost twenty years ago by anthropologist Daniel Miller (1988). He conducted interviews in a council estate in North London, which, though low-rise, was a high-density 'street in the air' and demonized like the other high-rises that we have discussed. For some residents, these dwellings were experienced as inhospitable, and they felt alienated from them. Moreover, Miller found that these residents were unwilling or unable to appropriate or change the material environment. For other residents, however, a potentially alienating environment became a site of belonging, evidenced by kitchen decorating and remodelling. Some would place colourful decorations such as tea towels, ornaments and curtains throughout the kitchen to draw attention away from the alienating features of the flat. Yet others completely remodelled and individualized their kitchens. Such initially alienating dwellings became home as they were personalized through material transformations.

Meanings of home are dependent on social, geographical and historical context, as are the material forms they take. One of the key geographical differences is that 'home' in continental Europe is predominantly associated with apartments. Whereas home became defined through the separation of public and private in Britain, in nineteenth-century Paris home was a place and an ideal that linked public and private and was manifested in the form of the apartment. In Paris of this time apartments were not only the most common building type, but they also most closely resonated with dominant ideals of home as a place for a family. The French word for house, 'maison', was stretched by architects and others of the time to also include multiple-occupancy rental buildings (Marcus 1999: 27). Home in Paris was defined through a melding rather than separation of public and private, at a number of levels. Home was to be located in, rather than away from, employment and commerce. Apartments, as a private building type, resembled public buildings in scale and use of materials. In both, building exteriors (public) and interiors (private) were similar rather than oppositional in style (28). Apartment interiors were then designed and represented in similar ways to the interiors of public places such as shops. As Sharon Marcus describes:

> Authors of pattern books frequently compared apartment buildings to stores, cafés, and theatres and recommended incorporating elements from commercial and civic buildings into apartment house décor ... [B]ecause café, restaurant and shop interiors used the same decorating principles and materials as domestic ones, commercial spaces often resembled apartments, but apartments open to public view and entry. The mirrors commonly placed behind plate-glass store windows reflected pedestrians just as looking glasses on salon walls mirrored people at home.
>
> (28)

One hundred years later, Norwegian homes took on the characteristics and practices ascribed to suburban homes though with one critical difference: they occurred in apartments rather than detached houses. A 1980s ethnography of working-class Norwegian women (Gullestad 1984) described practices in home as a setting for family and social life; production and maintenance of boundaries between home and the 'outside'; public and private spaces inside; and the ongoing (and gendered) decoration and transformations of the apartment. And in the kitchens of these apartments women forged relations with each other, cemented definitions of what was 'homely' and appropriate, and made social distinctions between classes. In a different national context, then, apartments became home.

SOCIAL RELATIONS OF IDEAL HOMES

Ideal homes not only assume a specific built form but they also encompass specific social relations. We outline these relations here, focusing on family, gender, class, race, age and sexuality. Our discussion is far from comprehensive. Its purpose is rather to illustrate the multi-scalar nature of home and the reproduction, contestation and reworking of ideal homes through home-making practices.

Home and family: gender, sexuality and age

Ideal homes embody familial-based gender relations. Imaginings of suburban homes and home-making practices within them position women as mothers and as primarily responsible for the domestic sphere. Real estate advertising provides many examples of this, none more blatant than those discussed by Louise Johnson (1993). In advertisements for new houses on Melbourne's fringe, photographs in house brochures depicted women firstly as homemaker – as responsible for and concerned with the kitchen and other sites of domestic work such as the laundry. Women were secondly depicted as seductress – lying on beds or enticing the viewer into other rooms. Men, on the other hand, were constructed as interested in games rooms, garages and backyards. Home-making practices, in particular domestic labour and mothering, simultaneously cement, contest and spatially extend this gendered vision of home (see Box 3.6 on Tupperware and Figure 3.3). In simple terms, the gendered domestic division of labour sees women taking primary responsibility for the day-to-day running of home and the creation of a 'home-like' environment. Women are far more likely to cook, clean and care for children, as well as manage the everyday running of the household. Men, on the other hand, are responsible for outside, and typically take charge of do-it-yourself projects. Absent from these constructions of ideal homes is acknowledgement of domestic violence and its implications for women's experiences of dwelling, as we expand in the next section.

Box 3.6   TUPPERWARE

The first Tupperware object was produced in the United States by the Tupper Corporation, headed by Earl Silas Tupper, in 1939 (Clarke 1997; Clarke 2001a). By the 1950s, Tupperware had become an icon of American suburbia and was an object of mass domestic consumption imbued with particular ideas about femininity. As Alison Clarke writes, 'Its fashionable pastel designs and amiable hostesses embodied the burgeoning aspirations of white, American suburbia' (1997: 133). Not only was Tupperware a material object that was in common use in American homes, but it was also sold by women through Tupperware parties that they hosted at home (see Figure 3.3). By 1954, 20,000 women belonged to the Tupperware party network in the United States as dealers, distributors and managers (Clarke 1997). Clarke argues that Tupperware parties 'sanctioned all-female gatherings outside the family. ... Whilst the pretext of the gatherings was domestic this did not preclude women from directing the conversation and interaction towards other concerns' (1997: 145). Although Tupperware parties might appear to have reinforced traditional and conservative feminine roles, Clarke argues that they also provided 'a pragmatic, proactive alternative to domestic subordination' (1997: 145).

Figure 3.3   Selling Tupperware at home. Reproduced by permission Ann and Thomas Damigella Collection, Archives Center, National Museum of American History, Behring Center, Smithsonian Institution.

Familial inscriptions of gender also constitute many apartments, though often through absence rather than presence. Developers of new apartments in Melbourne, Australia, envisage non-familial gender identities for women occupants (Fincher 2004). They speak of women without children, for whom home is part of a broader lifestyle that values the city and carrying out domestic activities, particularly eating, in public. These are sites, in other words, that position women more broadly than as mothers, and as carrying out domestic activities beyond the walls of the apartment. These gendered representations nonetheless remain narrow. They ignore the many apartments that are inhabited by women – young or old – who live alone. These gendered practices are always subject to contestation and negotiation.

The material geographies of suburban homes both accord with and contest family-centredness and familial gender relations. The choice and placement of objects such as furniture can be part of making houses family homes. Furniture inherited from, or given by, relatives can become permanent symbols of family. Similarly, the work of sorting, storing and displaying family photographs is often done by women in the enactment of their identities as mothers. According to Gillian Rose (2003), the use of family photographs enabled women to create a sense of 'homeliness', a material environment that felt 'homely'. But photographs also served another purpose, enabling the women to stretch domestic space beyond the home. The viewing and display of family photos connected them to family members and friends in other places and in the past (see Figure 3.4). Neighbourhood-based mothering practices entail a similar stretching of home, though in this case to the scale of the neighbourhood (Dyck 1990). The sharing of domestic resources such as childcare can transform the street into a home as well as creating safer neighbourhoods for children and less work for women. Indeed, for James and Nancy Duncan, one of the defining characteristics of suburban homes is that the suburb is home (2004). In image and practice, the suburban neighbourhood takes on the characteristics of home: private rather than public, a haven for the middle class, feminine (the sphere of women) and socially exclusionary.

Gender means a focus on masculinity as well as femininity, but there is a paucity of research on masculinity and home. John Tosh's (1999) historical work shows us that the middle class Victorian home was an important site in shaping bourgeois masculinities. Today, although active engagement with the cooking, cleaning and childrearing aspects of domestic labour are less common for men than women, the work of maintaining and fixing home most often falls

Figure 3.4   Materializing family at home. Photograph by Robyn Dowling.

to men. We also know that for men the spatial and emotional separateness of home and work is often strong (for example, Massey 1995a). Doreen Massey's research on men in the high-tech sector in Cambridge further showed that workplaces fulfilled the leisure functions of home for some of these workers (Massey 1995a). Finally, the focus in many societies on the breadwinner/provider role for men makes it difficult for them to feel comfortable at home during unemployment (McDowell 2000; Peel 2003) or retirement. Ann Varley and Maribel Blasco's (2001) research in urban Mexico suggests that for older men who have spent most of their adult lives at work, home is unfamiliar. They feel like a burden at home, find no space to relax in their own home and struggle to define a new role for themselves at home (130).

Varley and Blasco's research on older men leads us to the more general point that imaginaries and desires for home change as people grow older. The goal of 'ageing in place', in which older people remain in the family home, sees the meaning of home transformed. John Percival's (2002) study of the everyday uses of domestic spaces by older people in the UK is especially useful. Older people in their own home 'experience and assess domestic spaces with reference to three important evaluative criteria: do they facilitate routines, responsibilities and reflections?' (Percival 2002: 747). Home is used to promote a sense of continuity and self-determination, with potential self-determination associated with the ability to maintain upkeep of the home by themselves. Homes are sites of memory, filled with objects to remind them of family and events. Although older people don't often have children living at home, they continue to fulfil a parental role and prefer their home to have spaces for their adult children and grandchildren to be comfortable and stay overnight. Finally, Percival provides information to help us envisage the ways in which home (and in particular the dwelling) is multi-scalar for older people. On the one hand, specific parts of the dwelling – for instance the living room, or favourite armchair – become most homely and favoured spaces. On the other hand, for some older people, especially as they become less mobile, home becomes their world. Home may be a place from which to view the world as they sit at windows watching gardens and passers-by (745). Thus whilst home may become confined to the dwelling, it is a dwelling intimately connected to sites and relations beyond it. Yet not all older people live independently at home. In Mexico they also live in residential homes run by nuns, called *asilos* (see Hockey 1999 on residential homes for the elderly in the UK). These tend to be experienced in unhomely ways, as we outline in the final section of this chapter. For the men in Varley and Blasco's research, the institutional setting of the *asilo* was just as problematic as their family homes: '[i]ts heightened home-like characteristics make it unpopular with men, as they are unaccustomed to having their freedom curtailed' (Varley and Blasco 2001: 133).

Running through notions of ideal homes as family-centred is heterosexism: suburban homes, for example, are not imagined as the domain of same-sex couples or singles (Costello and Hodge 1999). Heterosexist practices and homophobia sustain these norms. For some gay men and lesbians visible signs of 'gayness' in their houses are removed, whilst the stereotype of gay men (most recently found in the television show *Queer Eye for the Straight Guy*) locates them in inner-city rather than suburban homes. Suburban homes can be queered, or made gay and lesbian. Andrew Gorman-Murray (in press, and in Research Box 4) points out how gay men reinterpret suburban spaces in queer ways. The idea of the suburban home can also be queered, or the idea questioned, turned upside down. Gorman-Murray shows that 'home' for gay men and lesbians can occupy many sites. For some, it may be the 'unhomely' beats and bars that create senses of belonging and intimacy beyond the dwelling, whilst for others neighbourhoods or friends' houses perform a similar function. For young gay men and lesbians living in the parental home, control can be experienced through the imposition of heterosexual norms and assumptions. Gill Valentine (1993) documents the ways in which the home is not necessarily a haven for young gay men and lesbians. Indeed, for some, home is a closet, a place where their sexuality is actively opposed and denied by their parents, making the home a place in which they are not comfortable and are unable to feel at home.

Children are presumed to be key inhabitants of 'homely' homes, though it is rare for children to be given any agency in the running or representation of these homes. Suburban homes are spaces of parental control and are spatially demarcated along generational lines. Wood and Beck's detailed analysis of the cultural, social and material minutiae of Wood's living room confirms that 'What is home for a child but a field of rules? From the moment he rouses into consciousness each morning, it is a consciousness of what he must and must not do' (1994: 1). 'Home rules' are made visible through their communication to children, and children's collaboration with these rules. The zoning of houses into child and adult zones is mentioned in Box 3.4 and can be amplified through home-making practices. In recent research conducted by Robyn Dowling in suburban Sydney homes, for example, in one home the garage was transformed into children's space. This was more than a playroom: it was also where the children ate and watched television. In this house, children were not even welcome in 'family' spaces.

Cross-national comparisons paint a more varied picture of the scale, or spatial extent, of home-space for children. In the Sudan, where the quality of dwelling structures and standards of living are poor, children call quite a large area home. They are free to roam independently for long distances from where they live (Katz 1993; 2005). In New York, children's home-space is increasingly confined to the house. In the middle of the twentieth century, neighbourhoods in places such as the UK, United States and Australia were home-spaces for children in that they were deemed safe for independent play. The late twentieth century saw increasing concern about the safety of streets and the competencies of children, alongside a devaluing of unstructured time and play (Valentine 1997). The scale of home for many western children is increasingly the house rather than neighbourhood, although for some children the home is a place of fear, violence and abuse. Moreover, it is the inside of suburban homes that have become the primary play spaces for children, due to the combined effect of the declining size of backyards, the increasing popularity of indoor activities such as computer and video games (McNamee 1998), and fears about 'stranger danger' beyond the home (Valentine

2001b). The suburban home is thus just as complicated a space for children as it is for adults, potentially both a homely and unhomely place.

Home, 'race' and class

As well as being presumed to be based around the heterosexual nuclear family and its associated gender relations, normative notions of home are dependent on particular classed and racialized imaginaries. In particular, home in these imaginaries is one that is white and middle class. We suggested in Chapter 1 that the idea of home as haven and a separate, private, sphere does not accord with many African-American experiences of home. It is also the case that these normative notions of home have underpinned direct and indirect state policies and economic processes that limit people of colour's access to home ownership and residence in a house considered to be homely. In apartheid-era South Africa, for example, urban policy aggressively expelled Africans from the city, destroyed established informal settlements and presented numerous other obstacles to African attempts to establish home (see Lee 2005). Less overtly, urban policies and practices such as mortgage lending in Australia and the United States, for example, have been similarly racialized and restrictive, as we outlined in the first section of this chapter.

Without denying the racialized underpinnings of these ideals and their material forms, a critical geography of home also draws our attention to resistances to, and transformations of, such ideals. In particular, scholars in the United States have recently pointed out that the location and imaginary of the suburban home was, and is, at least in part shared by African Americans. By 1940, almost one-fifth of African Americans in the metropolitan areas of the north and west lived in suburbs (Wiese 1999: 1596). The seeking and creation of this suburban dream was remarkably similar: people were propelled by the desire for a better life, open space, fresh air and home ownership. According to one resident, 'I got five rooms; they all got heat from an oil furnace. I got an electric stove and hot and cold running water from my well – and it's all paid for' (Wiese 1999: 1495). Home-making practices, and what constituted a homely home, were nonetheless subtly different. These differences included a heightened importance of home provisioning (through gardening and so forth, also highlighted in studies of working-class suburban homes such as Nicolaides 1999) and porous worlds of home and work, especially for women. In essence, parts of the United States saw the implementation of a 'working-class African American suburban dream' (Wiese 1999: 1502).

A recent South African case study by Rebekah Lee sheds further light on the material racialization of home (2005). In the apartheid era, housing for black South Africans was mass produced and part of processes of social control. Despite mass production, and the forced mobility of much of the black population, Africans 'grew' and transformed their houses, and created home 'out of the concrete and brick "shells" that they had received' (613). Drawing on interviews with two generations of African women in Cape Town, Lee suggests that families' home-improvement endeavours were an attempt to forge a more settled urban existence. Immediately upon moving in (typically in the 1970s), families plastered unfinished brick walls, added ceilings, laid flooring (such as wooden blocks or tiles) and installed internal doors (618) to ensure that they lived in a respectable house rather than what were described as the stables provided by the state. Later renovations in the 1990s were more substantial, reflecting a greater sense security of tenure. The large backyards were used to add additional backyard structures or extensions to the house itself to create a sense of 'spaciousness' (621). In so doing, 'their homes became a marker of the urbanising ambitions of first-generation women [the first occupants of the houses], and a testament to their rootedness in the city' (629).

Class is another social relation that is reproduced and contested within the home. Suburban homes are assumed to embody middle-class cultural ideals, such as home ownership and its signifiers of material achievement and stability (Gurney 1999a). Like the other social identities we have considered in this chapter, these class ideals are reproduced and contested through home-making practices. Middle-class identity is not a natural attribute of suburban homes, but a constructed and practised one. For example, the connections between suburban homes and an individual's class identity can be found in advertising. In the 1980s, the class position of Sydney houses was signalled and prioritized by housing being named after prestigious suburbs (Mee 1993). Suburban homeowners maintain the exclusivity of their spaces through shared practices. In Vancouver, shared notions of what grasses and plants were appropriate in front gardens worked to present a homogeneous and manicured external appearance that was felt to correspond to the 'classiness' of the homeowners. If an individual deviated from these norms, they would be spoken to by other residents (Dowling 1998b).

Home-making practices can also transform class identities. Levittown, New York, is widely recognized as one of the first mass-produced, large-scale suburbs in the United States (see Kelly 1993). William Levitt built the houses of Levittown using Fordist production methods, including prefabrication of

significant parts of the house. Levitt also built the houses for 'working families', consequently they were small and simply designed. As the size of inhabitants' families grew, and as they became more financially secure, houses
were dramatically transformed by their inhabitants. Garages, rooms and sheds
were added, kitchens and living areas redesigned, and attics converted into
bedrooms. But residents were doing more than enlarging their homes. In
customizing their homes, residents created 'the combination of picturesque
variety and community harmony that was the hallmark of the nineteenth-
century suburb' (Kelly 1993: 17). In other words, the material transformations cemented the middle-class nature of Levittown.

The middle-class ideals of suburban homes have also been used in public-
policy attempts to reconstruct working-class life. The living environments of
poor and working-class people have often been deemed 'unhomely', or
morally and aesthetically inappropriate and sometimes inadequate. The
suburban home, on the other hand, is constructed as the proper place of
moral, aesthetic and familial acceptability, even superiority. As a result,
suburban house design has been used in a number of different contexts with
the intention of 'reforming' working-class habits. This is especially the case in
council- or state-provided housing in Britain and Australia in the middle of
the twentieth century (see Turkington 1999; Bryson and Winter 1999). It was
thought that if the structure of the working-class dwelling changed, then
middle-class values of respectability and moral authority would follow. For
instance, the presence of a parlour or formal living space was supposed to
encourage working-class families to separate public and private activities. In
Australia, a home with a front garden was presumed to reproduce a similar
boundary between the public life of the street and the private world of the
family. In inhabiting the suburban house form, working-class residents both
contested and reproduced these ideals of home. In suburban Australia, public-
housing tenants used their parlour as additional bedrooms. Pride and care
were taken in front gardens (as depicted in Figure 3.5), in much the same way
as the residents of suburban Vancouver described earlier. Kathy Mee's recol-
lections of these public housing suburban homes are telling here:

> In many ways my childhood was characterized by a very ordinary living
> out of suburban life. My father was the household's wage earner, my
> mother worked in the home. My mother worked very hard on keeping up
> with contemporary standards of being a homemaker. Looking at slides
> from my childhood reminds me of how neat our house always was, and

Figure 3.5 Suburban front garden. Reproduced courtesy of Kathleen Mee.

mum's constant work of vacuuming, dusting, washing, ironing and so on to live up to those standards. When my father was home he occupied the male spaces of the house, the garden and the shed. Like many of the other men of Green Valley he created home through his 'handyman' labour, constantly doing up the house. The work of home making also involved the personalization of our space; the walls were covered with pictures and ornaments which marked the space 'our home'.

(Dowling and Mee 2000: 286)

Idealized views of home, in summary, are important to a geographical understanding of home. It is the suburban house that is most often presumed to be home because it is the form of dwelling that is seen to most clearly correspond to dominant meanings of home as haven from the public world, a space of familial intimacy and a site of domestic comfort (though Box 3.7 provides instances of alternative ideals). In this section we have illustrated how these notions of home inform the making of suburban houses and are often reproduced in the daily home-making practices of inhabitants. But a critical geography of home also alerts us to the power relations enacted and resisted through these material manifestations of ideas of home. Thus the suburban home becomes a tool used in attempts to modify working-class senses of home, uses of home and home-making practices. And suburban homes also demonstrate the oppressiveness of dominant ideologies of home, especially in terms of gender, sexuality and race. Home-making practices potentially contest and rework such exclusionary ideals of home in complex ways.

Box 3.7   SOCIALLY AND ENVIRONMENTALLY SUSTAINABLE SUBURBAN HOMES

Recently environmental critiques of suburban homes and home-making practices have come to the fore (also see Research Box 5 by Louise Crabtree on body/home/housing). The continued building of suburban homes is characterized as environmentally unsustainable. The large size of contemporary suburban houses gives rise to substantial use of non-renewable resources such as water, oil and gas. Energy use is also amplified by increasing standards of hygiene, declining tolerance for uncomfortable heat or cold, desires to keep nature 'outside' the home, and the priority given to comfort, cleanliness and convenience over environmental sustainability (see Kaika 2004; Shove 2003). Alongside the feminist critiques outlined in Box 3.4, concerns have arisen as to the social sustainability of suburban housing, especially because of its limited affordability and its reliance on ownership. In particular, the high cost of suburban housing limits its affordability to a small segment of the population.

A number of responses to these challenges have been made, many from the non-profit sector. In the UK, the work of the In Suburbia National Partnership has put suburban sustainability and regeneration onto the national agenda (www.civictrust.org.uk). As the Partnership explains, although 86 per cent of England's population lives in suburban homes, suburbs are missing from debates about the urban renaissance:

> The term 'suburbia' conjures up images of leafy streets which are pleasant and safe areas in which to live. This image masks the fact that many suburban areas are experiencing problems of decline such as run-down shopping centres, poor public

transport links, loss of local employment, some poorly performing schools and a rise in crime.

(In Suburbia Report)

Other responses focus on domestic architecture. In the 1980s a group of New York designers proposed to bring suburban and city homes together in order to balance the desire for single-family housing with urban land squeezes. Called Highrise of Homes, the project was to construct a ten-storey building where each floor would be a small village with interior streets and a number of houses, to 'allow personal identity in the cityscape, along with garden space in a multi-storey building' (Wakeley 2003: 147). But a survey of New Yorkers conducted by the designers found the proposal was not very popular, with reported comments such as 'hideous', 'vulgar'; and 'leave suburban housing in the suburbs' (Wakeley 2003: 147). The Peabody Trust, which is one of London's largest housing associations, with almost 20,000 homes, is a leading force in designing sustainable housing (www.peabody.org.uk). BedZED (Beddington Zero Energy Development), for example, is a mixed-use development in south London with 82 homes and 1,600 square metres of workspace, which is designed to be energy-efficient, eco-friendly, and has been built using recycled, reclaimed and renewable building materials.

UNHOMELY HOMES

In Chapter 1 we suggested that thinking about homeliness and unhomeliness is a powerful way of understanding the simultaneity of feelings of belonging and alienation associated with home. We also suggested that apparently unhomely places may be experienced in homely ways, and that its counterpart – unhomely experiences of places normatively defined as home – is also possible. Home, in other words, can be created, and takes different forms, in unlikely dwellings: student accommodation (Kenyon 1999); travelling caravans and mobile homes (see Box 3.8), residential homes for the elderly (see Hockey 1999 and discussion beginning on p. 113), and squatter settlements. Senses of home are nonetheless limited in such spaces. In their research in Ontario, Canada, for example, Erin Mifflin and Robert Wilton found that rooming houses had limited capacities to become home for tenants as 'rooming arrangements often failed to provide many of the characteristics – shelter, privacy, control, continuity, etc. – associated with the ideal of home' (Mifflin and Wilton 2005: 418). In this section we use three situations to illustrate unhomely homes, and the ways in which various home-making practices unsettle such assumptions about their unhomeliness: disability, housing and the meaning of home; domestic violence; and homelessness.

Box 3.8   HOMELINESS AND UNHOMELINESS IN MOBILE HOMES

Mobile homes – houses that are transportable and hence able to be located virtually anywhere – are an increasingly prevalent form of dwelling and location for home-making practices, especially in the United States. In the 1990s, mobile homes provided shelter for nearly thirteen million Americans and comprised a quarter of all new housing produced (Wallis 1997: vi). Also called caravans or trailers, this dwelling form, like the apartment, nicely illustrates the role of state regulation, home-making practices and limiting notions of what home should be in the creation of a particular form of house-as-home.

Mobile homes first became popular as vacation vehicles or 'tent trailers' in the 1930s. They became less mobile during the Second World War, when the United States government used house trailers (approximately 200,000) to house the many workers required in defence production centres (Wallis 1997: 90) and to house the workers in nuclear facilities in the postwar period. The latter half of the twentieth century saw mobile homes increase in popularity, with these 'dwellings that happened to be mobile' (Wallis 1997: 133) located on private land as well as in mobile-home parks, resorts and retirement communities. Despite this popularity, the idea that a mobile home cannot be home persists, in negative portrayals of residents of mobile homes as inferior 'trailer trash', and the perceptions of mobile homes as an inferior dwelling form because they are not permanent. Policies of governments and businesses exacerbate these perceptions of unhomeliness. In many cities a special zoning variance is required to site a mobile home in a residential district, and it wasn't until the 1980s that bank financing was available to purchasers of mobile homes.

Despite images of unhomeliness, research on living in mobile homes demonstrates the myriad ways in which they are perceived to be, and created as, ideal homes. Detached rather than built-in furniture is used to individualize the space, design changes include passageways to provide private bedrooms, and on-site additions are made to create larger, more comfortable spaces. For residents, a mobile home can be an ideal home, especially when compared to an apartment. As one resident interviewed by Wallis put it (1997: 187):

> We've been in nice apartments, though temporarily, but it's not the same. I've always had a little lost feeling when I go into one. It's not my home. It's not my furniture and just try and keep a child's hand from things. No freedom there ... neither we nor our child are any more restricted in a nice trailer court than we would be in the close housing of the cities or suburbs.

Similar homely practices have been described for European 'caravanners', with their dwellings and practices a 'miniaturisation of home' (Southerton et al. 2001).

Disability, housing and the meaning of home

Housing studies, research on domestic design and cultural studies of home have rarely focused on the implications of bodily impairment for the materialities and imaginaries of home. A body of recent research, including a special issue of Housing Studies on 'Housing quality, disability and domesticity' (2004), shows that

> For many disabled people, the physical design of dwellings is not well suited to their needs for access into, and ease of movement about and use of, domestic spaces. Rather most domestic design is premised on the production of dwelling spaces to facilitate use by people without bodily impairment.
>
> (Imrie 2004a: 685; also see Imrie 2005)

For many people, bodily impairment entails feeling out of place at home, and home is not experienced as homely or a welcoming, personalized space. One of the reasons for this, Rob Imrie (2004b) argues, is that house design generally reflects particular assumptions about corporeality that do not take impairment, disease and illness into account. For example, the standard height of kitchen units is based on someone standing up rather than in a wheelchair, and the 'spatial separation of the bathroom and toilet (upstairs) from daily living functions (downstairs) is premised on a walking and mobile person' (751).

It is not only design that makes home seem unhomely for people with disabilities. Imrie (2004b) also shows that the home can be a confining and isolating space, and that the reliance on carers for many people signals a loss of independence and personal control within the home. Moreover, because of the links between a person's feelings about, and experiences of, the home with corporeality:

> there are tensions between ideal conceptions of the home and the mate-
> rial, lived, domestic realities of disabled people. While aspects of the
> home may well provide for privacy, sanctuary, security and other aspects
> of 'ideal' domestic habitation, such provisions are always conditional,
> contingent, never secure, and likely to be challenged by, amongst other
> things, the onset and development of bodily impairment.
>
> (Imrie 2004b: 760)

Some participants in Imrie's research felt 'homebound' (753) or experienced home as a place of dependence and reliance on others rather than independence. Research on the experiences of children with disabilities underscores such experiences of unhomeliness. Interviews by Christine Oldman and Bryony Beresford (2000) with parents and children show alienating experiences of home. One of their respondents simply said 'We haven't got a home' (437), whilst another felt imprisoned in her home with two young children with learning disabilities. Children with limited mobility spoke of 'falling, tripping and banging themselves, and not liking the parts of the house where these sorts of accidents occurred' (437).

But home can also be a space of adaptation and transformation, whereby disabled people 'are not passive victims of insensitive design, nor necessarily resigned to dependence on others to facilitate aspects of their home lives' (Imrie 2004b: 756). Kitchens can be creatively adapted, floor surfaces changed to accommodate wheelchairs, furniture sold in order to 'de-clutter' the house, and internal walls knocked down to create larger internal spaces. One family had a new house built to suit the needs of their child, which opened up a new perception of home:

> If you've got your home right you can cope. This house is like a cocoon. It doesn't matter what's coming to us now. How can you make a tough decision in a house that's not home? Within 24 hours of being in this house it was like WOW! She was a different child. Her confidence increased over night. I can't describe to you the difference in Debbie.
>
> (Oldman and Beresford 2000: 439)

Such alterations can be costly, and it may be that unhomely experiences are magnified for the many people with disabilities on low incomes or living in housing that they are unable to modify.

Domestic violence

The home can be a place of violence and abuse as well as comfort and security, particularly for women. As Rachel Pain writes, 'The vast majority of incidents of violence against women take place in the home or other private and semi-private spaces. An accurate map of urban rape would highlight far more bedrooms than alleyways and parks' (Pain 1997: 233). According to Women's Aid in the UK, domestic violence is 'physical, psychological, sexual or financial violence that takes place within an intimate or family-type relationship and forms a pattern of coercive and controlling behaviour' (www.womensaid.org.uk). In other words, although domestic violence often takes place within the home, it is defined in terms of the relationship between people rather than its location although, as we discuss in Chapter 5, domestic workers also often suffer violence and abuse within the home. The British Crime Survey, 2001/2, reported 635,000 incidents of domestic violence in England and Wales. Women made up 81 per cent of the victims and 57 per cent of victims – a rate higher than any other crime – had been involved in more than one incident. Fewer than 35 per cent of incidents were reported to the police (www.womensaid.org.uk). There are about 400 shelters for women in Britain, and the first shelter for male victims of domestic violence opened in early 2004. Men can suffer domestic violence in both heterosexual and same-sex relationships but they are less likely than women to have been repeat victims of domestic assault, are less likely to be seriously injured, and are less likely to report feeling fearful in their own homes (www.womensaid.org.uk).

Domestic violence is also a significant cause of homelessness for women. According to Shelter in 2002, 40 per cent of all homeless women stated that domestic violence contributed to their homelessness (www.womensaid.org. uk). Domestic violence is not only a significant cause of homelessness for women, but is also experienced by homeless and insecurely housed women. Paula Meth (2003) argues that understandings of the home in relation to domestic violence need to be much more nuanced and differentiated. In her research on domestic violence and homeless and insecurely housed women in Durban, South Africa, Meth critiques three assumptions about the home that are prevalent in research on domestic violence: first, that 'the home is often assumed to be a formal material entity' in studies of domestic violence; second, that 'the home is also assumed to be a space of privacy offering separation from an outside public'; and, third, that 'the "domus" identified in domestic violence studies is assumed to symbolize the home for its residents' (317). For the women that Meth interviewed, '[t]heir realities of vulnerability, insecurity, exposure and informality suggest that their experiences of domestic violence cannot be examined using the frameworks adopted in western studies' (327). Questions that researchers often ask when studying domestic violence in western contexts – such as 'Did you lock the door? Did you hide somewhere? Did you lock the children away? ... Did you gain access to a shelter? Did they help re-house you?' (357) – often make assumptions about the space of the home that are not applicable to homeless and insecurely housed women. Reflecting on the ways in which the spaces of the house and the bedroom are often used in uncritical ways in studies of domestic violence, Meth poses other important questions that should be considered: 'Are these formal houses, and are they permanent structures. Are the bedrooms really bedrooms or simply spaces behind curtains within a single shack? Do they have locks on the doors? Are there doors?' (319).

Homelessness

Homelessness is another important example of assumed unhomeliness. Homelessness is not simply the absence of home, but overlapping feelings and constructions of being at home and not at home. One aspect of homelessness is 'houselessness': of 'being without access to shelter or relying upon institutional places like shelters to sleep' (Kellett and Moore 2003: 126). Though collecting statistics on houselessness is obviously difficult, it is estimated that 600,000 people are houseless in the United States on any one night, 1.5 million in South Africa and 18.5 million in India (Kellett and Moore 2003: 125). The causes of homelessness are complex. These include, as outlined by Jennifer Wolch and Michael Dear (1993), economic factors such as deindustrialization and relative decline of minimum wages, the political restructuring of the welfare state (especially the deinstitutionalization of the mentally ill and declining state provision of affordable housing), the increasing lack of affordable housing in cities, and other factors such as domestic violence, particularly for women (see p. 126). Homelessness is also a direct result of disasters such as earthquakes, floods and hurricanes. For example, an estimated 2.5 million people were made homeless by the earthquake in northern Pakistan in October 2005 (*Guardian*, 11 October 2005), and, six months after the Asian tsunami in December 2004, half a million people remained homeless in Sri Lanka alone (International Federation of Red Cross and Red Crescent Societies: www.ifrc.org). Many other people are forcibly evicted from their homes (see Box 4.6 on domicide). The consequences of homelessness are far-reaching, as Wolch and Dear explain:

> [c]oping without a home is a stressful, time-consuming occupation. Every single day, homeless people are faced with the task of securing food, shelter, and other necessities of life. Frequently they are obliged to negotiate complex bureaucratic systems, endure alienating and dehumanizing service-delivery routines, and risk arrest or jail. They live with the physical and psychological consequences of poor diet, inadequate rest, and lack of health care.
>
> (Wolch and Dear 1993: 246)

But houselessness and homelessness are not the same. It is possible to be homeless even while physically sheltered. Children growing up in unsuitable foster care, or in abusive domestic contexts, certainly do not label their environment home even though they have a roof over their head. For some of these children, leaving these abusive family situations may lead to them becoming houseless. Catherine Robinson (2002) gives us the example of teenagers Josie and Andy in Sydney. Having previously been 'houseless', Josie and Andy currently live in a two-bedroom council flat. They do not, however, consider it home, but see it as a site of violence and fear. People to whom they owe money visit and threaten them; relatives drop in and do not leave for many weeks.

Recognizing that houselessness and homelessness are not the same, the United Nations has recently defined homelessness as 'a condition of detachment from society characterized by the lack of affiliative bonds … [that] carries implications of belonging nowhere rather than having nowhere to sleep' (UNCHS/Habitat 2000, cited in Kellett and Moore 2003). No matter how well intentioned, such a definition ends up portraying homelessness 'as a totalizing condition of lack' (Robinson 2002: 31). Being homeless means having no shelter, no community ties, no sense of self. A critical geography of home navigates a middle path through these polar definitions of simply being without shelter and being without any sense of belonging or identity. As April Veness (1993) suggests, homelessness is an 'unhome': attempts to create home in situations and spaces not usually amenable to home-making processes. Three examples illustrate this.

In the United States, living in inadequate and substandard housing would define someone as homeless. But what is inadequate and substandard is politically and culturally embedded. It may include, for example, people without access to running water and electricity (Veness 1993). But as Veness shows, this definition comes from one segment of society that is imposed on others. Moreover, it is dependent on a definition of home that does not mesh with the individual realities of many poor people's lives (Veness 1993: 322). For example, Veness describes the home of a woman whom she calls Miss Edda. Miss Edda's home in rural Delaware did not meet state-sanctioned definitions of home, and is worth describing in some detail:

> In one mobile home she slept, in another she cooked and stored her clothes, and in the shanty she housed assorted personal belongings. There seemed to be no identifiable 'hearth' to her home. Scattered about Miss Edda's 'living room' were: several abandoned cars (one that served as a chicken coop); several refrigerators in differing states of repair; an old lawn mower and washing machine that she used; various sofas and chairs; and random piles of post-incinerated bottles, cans and plastic. From the utility pole at the edge of the property, a sagging electrical cable brought power to one of the mobile homes. A set of extension cords stretched across the grass linked the mobile home to whatever working appliances Miss Edda wanted to use outdoors. Miss Edda had no telephone, no longer drove a car, drew her water from a hand-pump and walked past the garden, beyond the pigsty, to an outdoor privy.
>
> (Veness 1993: 328)

This was home for Miss Edda, a place with which she could feel satisfied and in control.

Homelessness is also equated with the space of the homeless shelter. Like other institutional forms of dwelling, the homeless shelter is not conventionally described as home. The public permeates every aspect of the lives of inhabitants. There may be licensing regulations, for example, that forbid certain activities (such as smoking) and people (visitors) in individuals' rooms. Beyond the public and the private, Veness illustrates the assumptions of home written into the design of shelters. Shelter design became a social and architectural issue in the United States of the 1980s. Many new shelters were proposed, and these invariably took the form of the single family house described in the first section of this chapter. This design was adopted to create an 'ideal' home environment and remodel the behaviour of homeless people, even though it was not predicated on either the home-making practices or their ideal of home. For example, separate spaces for family and non-family, adults and children, were planned (Veness 1994). This 'model' home became an unhomely home through the imposition of rules and expectations. Thus in one shelter in urban Delaware, residents were not allowed to wear shoes in the living room, nor were they to share resources such as childcare (Veness 1994: 162). It was a home (though narrowly defined) in intent and design, but not a home because of the limitations placed on inhabitants' autonomy and freedom to use the space independently. Moreover, Veness shows us how homeless shelters assume homeless people have not previously had a home.

Our final way of thinking about homelessness as homely and unhomely is through the senses of home created by homeless people (see Box 3.8, which describes two recent *Big Issue* publications on home, one from the UK and one from South Africa). As Gill Valentine writes, 'Homeless people create relationships, social networks and appropriate spaces which take on many of the meanings of home (e.g. abode, identity, roots), which the homed attribute to conventional forms of housing' (Valentine 2001a: 101). Senses of home and belonging are created by those without shelter, in many and varied ways. For example, young homeless adults living in shelter accommodation in London described how the shelter offered them security, independence and freedom, and their interaction with others in the shelter created a sense of family. But this was also an unhomely home because this sense of belonging was always invaded by a sense that this home was neither permanent nor did it accord with their ideal of home (Kellett and Moore 2003: 133).

Box 3.9   HOME, HOMELESSNESS AND THE *BIG ISSUE*

Two recent publications from the *Big Issue* have focused on home (see www.bigissue.com and www.bigissue.org.za for more information). The first, *The Big Issue Book of Home* (Ephraums 2000), was published in Britain and includes photographs, poems and other writings on home. The second was a photographic issue of the *Big Issue* magazine on the theme of 'Home', published in South Africa in January 2005, which also includes essays and quotations about home. Both publications focus on the question 'what does home mean?' and feature a wide range of answers from *Big Issue* vendors and other writers and photographers through the juxtaposition of different possibilities of home in both text and photographs. Sometimes this juxtaposition works within the same image, as shown by a photograph by David Goldblatt of a hawker standing alongside a billboard advertising a luxury home in Johannesburg (Figure 3.6).

The two publications vividly convey the differences and similarities between experiences and meanings of home and homelessness in Britain and South Africa. Whilst both publications feature a wide range of homely and unhomely dwellings, including family homes, the pavement, a prison cell, and a travellers' site, the political context of home and homelessness is different in both places. As James Garner, the editor of the South African *Big Issue*, writes,

> many South Africans attach a sense of displacement to the idea of 'home.' For many *Big Issue* vendors, this sense of displacement stems from the fact that they have left their family homes in rural areas to set up new homes in cities, driven by a need to find work. In other cases it is because their families were forcibly removed from one area and relocated to another during the apartheid era, a legacy that remains with us as the land reclamations process slowly attempts to address the dislocations of the past.
>
> (Garner 2005)

In both publications, the home is described as a place of both fear and desire, and as a place to escape from and to escape to. For

Figure 3.6 'George Nkomo, hawker, Fourways, Johannesburg, 21 August 2002'. Photograph by David Goldblatt. Reproduced courtesy of David Goldblatt.

example, Geraldine Matilda Rhode, a *Big Issue* vendor in Cape Town, says 'I used to live in a flat with my children, but I couldn't pay the rent so now I'm on the street. I dream of a small room where my children and I can feel safe and free. I dream of a place of my own that I can call my home' (*Big Issue* 2005). In contrast, Fi Tillman writes about British streets as 'a different type of home ... the home that unites thousands of children every year': 'Do I long for a two up two down, central heating and hot water ... no, I long for those corroding memories to dissipate, for someone to accept me, just like the street does. Cold stone doesn't beg a plethora of answers; it just sits and accepts the burden without question' (Tillman 2000: 81).

CONCLUSIONS

This chapter has canvassed some of the ways in which ideas of home become attached to physical structures we call dwellings, paying particular attention to an imaginary of home that casts the social relations of middle-class, white, heterosexual, nuclear families, and its material manifestation in the form of the detached suburban house, as an ideal, or homely, home. But in casting a critical geographical eye over this ideal we have begun to dismantle its unquestioned acceptance. We have firstly drawn attention to the diverse institutions and individuals producing and reproducing the ideal. These dwellings are constructed (literally and figuratively) as ideal homes by the actions of individuals, businesses and the state and through their representation in popular culture. In this we have underscored the political significance of ideal homes, and their connections with, for example, nation-building goals, state policies on race and urban space, and land development policies. House-as-home is also political through its connections with power relations and consequent processes of struggle and resistance. Dominant imaginaries of home are always contested, sometimes reworked, sometimes reproduced, through home-making practices, and part of identity formation. We have secondly demonstrated the multiple scales of house-as-home. Suburban house design, and the ideal home that informs its production, confines home to house and separates public and private, whereas in apartment living public and private become more blurred. Simultaneously, suburban home-making practices, especially those surrounding parenting, extend home beyond the house to the suburb. For homeless people home can be the scale of the body. The multiple scales of both homely and unhomely homes bring us to our third and perhaps most important point of the chapter: that there is no neat distinction between the two. For purposes of exposition we divided this chapter into homes considered homely, and those considered unhomely. But it is the similarities across this division that are most startling. Suburban houses can be experienced in decidedly unhomely and alienating ways, such as in cases of domestic violence or by gay men, lesbians and bisexuals. And purportedly unhomely sites and experiences, such as mobile homes or dying at home, can and do become sites of affirmation and belonging. In the next two chapters we begin to mobilize the home beyond the scale of the dwelling, focusing first on home, nation and empire and then on transnational homes. In both cases we continue to unsettle the distinction between homely and unhomely homes.

Research box 3 FRIDGE SPACE: GEOGRAPHIES OF TECHNOLOGY AND DOMESTICITY

Helen Watkins

I work at home, where household chores, domestic objects and machines inhabit both my research and my daily life. My desk, chair, computer, kettle, coffee cup and fridge support and discipline my routines, yet such things are so familiar in the social worlds through which I move that they fade almost unnoticed into the fabric of people's homes and lives. Curious about their histories, functions, journeys and meanings, I try to peel away the patina of ordinariness that keeps mundane objects and technologies 'hidden' in plain view.

Long-standing interests in home, material culture, domestic practice, technology, classification and display have shaped my research. My MA analysed the notion of 'curating' home in Kettle's Yard, Cambridge (Watkins 1997). Its founder, insistent that art was better experienced in a domestic setting than the rarefied atmosphere of a museum, opened his home to visitors. In this 'home made-public' art works are displayed amongst natural objects and household furnishings, complicating public and private and tangling curating with everyday orderings of home. More recently, I participated in a study of sustainable domestic technologies, which focused on kitchens and bathrooms, practices of cooking, cleaning and caring, the normalization of resource-intensive appliances, and the reordering of routines.

My PhD explores geographies of the domestic refrigerator in Britain. I ask how refrigerators have shaped and been shaped by cooking, eating and provisioning practices and what role they play in constructing homes, bodies, domestic knowledges and gendered identities. Adopted for food preservation, then co-opted in unintended but now familiar ways as notice boards and display spaces, refrigerators, I argue, have been doubly domesticated (Watkins forthcoming). A byword for a site of communication and creativity where anyone can be an artist or a poet, there are even 'fridge door' webpages for sharing information, poetry, children's art or handy household hints. I suggest it is its domestic connotations that make 'the fridge door' an accessible and unintimidating space.

I am also interested in domestic objects elsewhere. I ask what happens when fridges fail or cease to fit their users' needs, where they travel and whether domesticity gets smuggled with them or reconfigured on the way. I follow cycles of repair, reuse, disposal, destruction and preservation as refrigerators circulate through repair shops, recycling facilities, dumps and crushers, warehouses and museums. Alongside these physical journeys are conceptual ones as fridges slip between categories, from kitchen appliance to historic artifact or hazardous waste.

I had not anticipated that this project would become as much about domesticity as refrigerators, but instances of evident discomfort at refrigerators straying 'out of place' prompted questions about locating or containing domesticity. One example takes Britain's 'fridge mountain' crisis. Triggered by tightened CFC-reduction legislation in 2002, disposal of redundant refrigerators stalled and a usually unobtrusive object shifted into hypervisibility. Another example turns to fridges in museums. I examine how appliances are ordered, exhibited and stored in The Science Museum in London and note initial reservations about whether 'domestic' technology had a legitimate place in a national museum of science and industry.

As one link in a far-reaching 'cold chain', refrigerators illustrate the folding, stretching and interweaving of diverse and distant times and spaces. They also demonstrate the 'reach' of home, and the interconnectivities of scale, by showing how micro(bial) geographies of bacterial decay and seemingly localized preoccupations with rancid butter or limp lettuce are simultaneously imbricated in global issues of resource use, ozone depletion or environmental change. Fridges have led me well beyond the home, but always back again to the kitchen, to domesticity, daily life and the scaffolding of things with which we live.

Helen Watkins is a PhD candidate in Geography at the University of British Columbia. Her research interests lie in domestic space, material culture, histories of technology, ordering practices and cultures of display. She recently worked on an ESRC project on sustainable domestic technologies at Manchester University (award number 332 25 007; principal investigator Dr Dale Southerton). Her PhD thesis explores cultural histories and geographies of the refrigerator in Britain, and is entitled 'Fridge Space: Journeys of the Domestic Refrigerator'.

Research box 4 QUEERING HOME, DOMESTICATING DEVIANCE
Andrew Gorman-Murray

My approach to studying the home is via dissident sexuality. Concepts of home are gaining wide currency in research into non-normative sexualities. Eng (1997: 32), for instance, asserts that 'anxieties about loss of home remain psychically central to queer cultural projects and social agendas'. Analysts of gay, lesbian, bisexual and queer (GLBQ) mobility have been particularly sensitive to the way sexual dissidents 'on the move' invoke ideas of home. Researchers of queer migration and tourism have evoked the notion of queer 'homelands' to which GLBQ subjects relocate or travel (Weston 1995; Waitt and Markwell 2006). Meanwhile those working with the heuristic of 'queer diaspora' have theorized the migration of sexual dissidents as a form of 'homecoming' (Fortier 2001).

Rather than looking at the imagined geographies of 'homelands' or 'homecomings', my project focuses on lived experiences, material constructions and social practices which make particular locations homes for GLBQ subjects: how do sexual dissidents *make* and *use* home in ways which resist heteronormativity and affirm sexual difference? In doing this I intend to bring together and contribute to geographies of both home and sexuality. Within geographies of sexuality, home has largely been conceived in a negative manner. The family home, for instance, is posited as a site of heteronormative socialization that alienates GLBQ identities, while sexual dissidents' residences are not private, but regulated by external heteronormative surveillance (Johnston and Valentine 1995). My project aims to reclaim the importance of home for deviant sexualities, demonstrating how certain social practices cultivate identity-affirming attachments to home. Geographies of home, meanwhile, have drawn on feminist literature to show how home is not only gendered, but also classed and raced, eliciting the overlapping identities that determine how homes are made and used. I intend to add to these layered identities of home by drawing attention to the way homes are also sexualized, suggesting that while home is discursively heteronormalized, sexual dissidents can and do create and use homes in ways which resist heteronormativity. This will make a critical contribution to debates around the interpenetration of space and sexuality, bringing forth the heretofore neglected importance of home in this mutually constituted relationship.

In this discussion I have been using the term 'home' rather loosely. Home is a multi-faceted concept, incorporating material, social and symbolic dimensions, and competing meanings, on scales from the domestic to the diasporic (Blunt and Varley 2004). Rather than making home an 'empty signifier', this looseness and flexibility is conceptually fruitful. Home is much more than a mate-rial site – it is a place constructed through social practices and personal attachments. This means that home, while having a 'real' material dimension, is not bound to a pre-ordained site, such as the physical house. Moreover, since the social practices, activities and relationships that make home are extra-domestic to begin with (Moss 1997), so too are the spaces that can be designated 'home'. These conceptualizations are helpful for understanding the relation-ship between sexual dissidents and home. Instead of separating the domestic environment from 'outside' space, I contend that the ways in which GLBQ subjects *use* residential spaces *stretches* home beyond the domestic and into external sites, and consequently certain public and private spaces and activities imbricate in their constructions of home.

To understand how sexual dissidents configure home and belonging through social practices and relationships, my research entails several discrete projects using both autobiographical and interview material. These include: the imbrication of domestic and public spaces in gay men's uses of homes; relocation and the search for home as a quest for identity; the role of supportive family homes in the constitution of confident gay/lesbian identities; gay/lesbian youth remapping home and belonging, incorporating multiple sites into a layering of identity-affirming attachments; home as an extension of embodied identity and bodily practices; the non-normative ways sexual dissidents adapt material home-spaces for unintended uses; how home is used to constitute the 'coupled' identities of GLBQ partners; and the interpenetration of house and neighbourhood in constructions of home. These inquiries will begin to generate a more explicit appreciation of the role and use of homes in the lives of sexual dissidents and in GLBQ cultural practices, contributing new understandings to geographies of both sexuality and home.

Andrew Gorman-Murray is a PhD candidate in Human Geography at Macquarie University, Sydney. His research examines the home-making practices of gay, lesbian, bisexual and queer (GLBQ) Australians. Entitled 'Queering Home, Domesticating Deviance: Geographies of Sexuality and Home in Australia', the research explores how GLBQ Australians find, make and use homes in various ways that resist heteronormativity and affirm sexual difference. Andrew has broader interests in feminist and poststructural approaches to space, place and identity, and has published in *Public History Review*, *Social and Cultural Geography* and *Traffic*.

Research box 5   BODY/HOME/HOUSING
Louise Crabtree

My research concerns the overlaps and interactions between sustainable housing, citizenship and affordability – but how to research this? If we start with a single human being we are already dealing with a mess. Humans leak, contradict, overlap and intertwine. Irigaray reminds us of our messiness, our porosity, our ability, desire and propensity to comingle. Mouffe informs us that we each contain wrestling multitudes, constructed and negotiated socially and individually, and that citizenship must embrace this. Morris tells us of embedded, open houses, places from which to go forth, not bound and gagged. Rhizomes break forth. At which point Irigaray distils the perceived male need to construct home in order to assert that elusive fixity: the public–private divide as a constructed boundary, nervously drawn tight around an anxious self struggling for control, certainty, Mother. So, what is a household? What is sustainability? What is citizenship? Or, what could they be? With these questions burning I set out to find housing projects that were attempting to deal with the dialogues between – or comingling of – home, self, environment and economy, and which were seeking justice in each of these realms.

However, all things contain multitudes and are embedded. This is not news. I could not write a thesis on the basis that the everything is connected to the everything else. So a level of filtration is required, some arbitrary way to herd the cats. I can distil things around the rhizome theme – groups, households, individuals as citizens – but isn't that a bit of a one-joke thesis? Or I can attack the need to seal things off – let's call it Reason, the dominant darling of the Enlightenment that believed it could master and muster. Getting over the binaries of modernity speaks nicely of edges and multiplicity; messy, oozing, multiple selves. Rhizomes and their edgy permacultural friends can be enlisted to this end; John Ralston Saul can come along for the ride.

With one quick twist, we have a range of options. The household can be seen as the affiliation of various multiple selves enacting a range of possible articulations. This body, as leaky as the rest of us, is itself enacting a range of possible articulations. Or could be. I want to explore ways of enhancing the ability of households and their messy humans to enact and extend these articulations, in ways that celebrate and enliven the ecological embeddedness of our actuality. Messy humans, dirty economies and leaky houses. Three pies in the face of Reason: splat, splat, splat.

What's that look like? An investigation of models such as cohousing, ecovillages, cooperatives (public, private, both), community land trusts, sweat equity, dual mortgages, with a view to working out how we can do more of them. Interview the bureaucrats and current masters of the mainstream housing universe – developers, planners, government providers. Go well beyond that universe by visiting sites and interviewing proponents of housing done in the name of redesign, rethink, reinterpret, reoccupy in post-consumerist moves to sort out this unaffordable, unsustainable, inequitable mess we're in. Look at the material needs of our daily lives (not GDP, not consumer addictions) and ways of access, engagement and delivery that free up our own dances rather than dictate imperialist waltzes. Shall we?

Louise Crabtree has been teaching and researching sustainability over the past eight years at the University of New South Wales and Macquarie University, where she is currently completing her PhD 'Messy Humans, Dirty Economies and Leaky Houses: Citizenship, Sustainability and Housing' in the Department of Human Geography. This involves documenting sustainability and affordability in

4

HOME, NATION AND EMPIRE

Two images produced for an English audience in 1857 encapsulate the connections between home, nation and empire that are the focus of this chapter. The first, entitled 'The sinews of Old England' (Figure 4.1), is a painting by George Elgar Hicks, and is a classic representation of the gendered spaces of home and nation. The married couple and their child are pictured on the threshold of their home, with the man – a muscular embodiment not only of manhood but also of the English nation – gazing beyond the frame to the world of physical work beyond. In contrast, his wife is portrayed as a domestic subject, responsible for the well-ordered domesticity that is visible beyond the threshold. The title of the painting implies that families like this, pictured on a pastoral threshold between masculine work and feminized domesticity, embody an idealized idea of English nationhood. The family home appears as an integral location for imagining the nation as home.

Contrast this painting with another image of the home from 1857. Entitled 'How the Mutiny came to English homes' (Figure 4.2), this drawing depicts the home as a site of fear and danger. Once again, the domestic home represents the English nation as home, as shown by the title of the drawing and the box on the chaise longue. But rather than depict the home and nation as an idealized site of peace, comfort and order, this image shows its violent disruption, as insurgents have forced their way over the threshold to threaten the British woman, her children, the home and, by implication, the nation itself.

Figure 4.1 'The Sinews of Old England' by George Elgar Hicks. Unknown private collection.

Figure 4.2 'How the Mutiny came to English Homes'. Source unknown.

Like Hicks' painting, the home is embodied by a young wife and mother, but in this case she is portrayed as vulnerable and defenceless, in part because her husband is absent apart from his portrait on the wall. Although the style and furnishings of the home appear to be quintessentially English, the drawing is located in India and depicts the start of the so-called 'Mutiny' of 1857–8. Both here and elsewhere, the severity of the threat posed by the uprising to British rule in India was depicted most graphically in terms of the vulnerability of British women and children and their homes (Tuson 1998; Blunt 2000b).

Both images show the symbolic importance of the domestic home to an idea of the nation as home. Whilst 'The sinews of Old England' represents an idealized home and family, the drawing depicts the vulnerability of English homes and families in the midst of a violent uprising. The ability of the latter drawing to convey the threat posed by the 'Mutiny' in 1857 was due, in large part, to the wide currency of idealized notions of home, family and nation as represented by Hicks' painting. Moreover, by depicting an English home in India, the drawing shows how the imaginative geographies of home, family and nation in the nineteenth century were inseparably bound to the wider empire.

This chapter explores the imaginative and material geographies of home in relation to the politics, lived experiences and conceptualizations of nation and empire. We argue that the home as a lived place and as a spatial imaginary has been mobilized and contested in ways that shape and reproduce the discourses, everyday practices and material cultures of nation and empire. Rather than view the home as a private space that remains separate and hidden from the public world of politics, we argue that the home itself is intensely political both in its internal relationships and through its interfaces with the wider world over domestic, national and imperial scales. We explore the critical geographies of home, nation and empire in three main contexts: imperial homes and home-making; homeland, nation and nationalist politics; and the politics of indigeneity, home and belonging. In each context, we argue that the home is a site of both power and resistance. The home on a domestic scale is intimately bound up with imperial, national and indigenous politics that are themselves articulated and contested through material and imaginative geographies of home.

A relational understanding of home underpins our concern here with the ways in which material and imaginative geographies of home – and the ways in which nations and empires have been conceptualized as home – are forged in relation to 'foreign' or 'unhomely' places and practices. In her research on postcolonial fictions of home, for example, Rosemary Marangoly George writes that 'homes and nations are defined in the instances of confrontation with what is considered "not home", with the foreign, with distance' (George 1996: 4). In similar terms, Amy Kaplan writes in her study of American imperialism and national identity that 'The idea of the nation as home … is inextricable from the political, economic, and cultural movements of empire, movements that both erect and unsettle the ever-shifting boundaries between the domestic and the foreign, between "at home" and "abroad"' (Kaplan 2002: 1). Kaplan points out that the term 'domestic' has a double meaning, referring both to the space of the nation and to the space of the household. Both of these meanings, in turn, are closely bound up with shifting ideas about the 'foreign'. Terms such as 'foreign' and 'domestic' are not neutral, but are rather 'heavily weighted metaphors imbued with racialized and gendered associations of home and family, outsiders and insiders, subjects and citizens' (3). We investigate the material effects of such metaphors by exploring the salience of the home in imperial, nationalist and indigenous politics.

This chapter thus extends our discussion in Chapter 3 about the home as a site of inclusion and exclusion beyond, as well as within, the domestic household. Like our elaboration of house-as-home in Chapter 3, and like our discussion of Figures 4.1 and 4.2, we are particularly concerned with the ways in which the intersections of home, nation and empire are gendered, racialized and underpinned by the assumed heterosexuality of home and family life. Throughout this chapter, we investigate the transnational mobility of home, both as lived experience and as a spatial imaginary, and the intensely political interplay of home and homeland in imagining the nation. These themes are also important in Chapter 5, where we turn to focus on transnational homes in the contemporary world.

IMPERIAL HOMES AND HOME-MAKING

The exercise of imperial power not only relied on imaginative geographies of 'other' places (Said 1978), but also on imaginative geographies of home within the metropolis, between the metropolis and the wider empire, and across a wide range of colonized spaces. As Amy Kaplan explains, 'domestic metaphors of national identity are intimately intertwined with renderings of the foreign and the alien, and … the notions of the domestic and the foreign mutually constitute one another in an imperial context' (2002: 4). Such imaginative geographies and domestic metaphors were closely bound up with the materialities of home. Domestic practices and ideas about the nation as home were not only materially reproduced and recast over imperial space, but were also shaped at home by the wider empire. We begin by charting the critical interplay of home, nation and empire through various sites of imperial domesticity both in the metropolis and in the wider empire, before focusing on two examples in more detail: frontier home-making and American nation-building and the contested place of British homes in imperial India.

Home, metropolis and empire

Within some studies of empire, the home is represented as a confining place from which to escape. Adventure beyond the home, for example, has been an important theme in work on imperial masculinities, as shown by John Tosh's work on the 'flight from domesticity' for British men in the late nineteenth century (1995) and Richard Phillips' study of the imperial spaces of adventure far from home in boys' fiction (1997). In addition, research on imperial sexualities has explored the licence to transgress away from home and for men to live with other men, rather than in a heterosexual home and family (Ballhatchet 1980; Duncan 2002). The spatial extent of the British Empire at this time also enabled British women to travel more widely than ever before. Some women travelled independently and shared in imperial power away from the feminized domesticity of life at home. Mary Kingsley, for example, travelled to West Africa in the 1890s once she was free from domestic respon- sibilities as her parents had both died and her brother was abroad. Although Kingsley's travels enabled her to transgress feminized domesticity when she was in West Africa, she was repositioned within an appropriately feminine domestic sphere whenever she returned to London, where she worked as her brother's housekeeper (Blunt 1994). Other women left home as domestic workers, often under the auspices of various emigration societies that orga- nized colonial settlement for unmarried working- and middle-class women (Kranidis 1999; Myers 2001; Pickles 2002a). Many Europeans and Americans lived overseas as missionaries, and sought to convert local people not only to Christianity but also to western domesticity (Comaroff and Comaroff 1992). But most metropolitan subjects who travelled in the empire did so to set up homes both with and for their families. The material practices of imperial home-making reproduced and reinscribed imaginative geographies of both nation and empire as home (see David 1999 for a helpful review of work on imperial domesticity).

Imperial home-making – like the effects of imperialism more generally – was not a one-way process, transported solely from the metropolis to the wider empire. Rather, metropolitan homes, and ideas about the nation as home, were themselves intimately shaped by imperial politics and encounters. In their study of British colonial missionary work in southern Africa, for example, Jean and John Comaroff write that

> the evidence suggests that colonialism itself, and especially colonial evangelism, played a vital part in the formation of modern domesticity *both* in Britain and overseas; that each became a model for, a mirror image of, the other; [and] that historians have underplayed the encounter with non-Europeans in the development of Western ideas of modernity in general, and of 'home' in particular.
>
> (Comaroff and Comaroff 1992: 39–40)

Maternity and domestic consumption in metropolitan homes were also shaped by imperialism. In a classic study, Anna Davin (1978) has revealed the extent to which imperial politics influenced practices and representations of motherhood in Victorian Britain, whereby the promotion of public health, hygiene and home economics explicitly tied domestic reproduction to national and imperial ideas about racial purity and strength. As Anne McClintock explains, 'Controlling women's sexuality, exalting maternity and breeding a virile race of empire-builders were widely perceived as the paramount means for controlling the health and wealth of the male imperial body politic' (1995: 47). Furthermore, the growth of commodity consumption during the nineteenth century also reflected and reproduced imperial domesticity and ideas about the nation and empire as home, as shown by the material cultures of textiles and food, and through their promotion and display in early advertising and department stores as well as through their use and consumption at home (Chaudhuri 1992; Richards 1990; McClintock 1995; Zlotnick 1995; see Research Box 6 by Sarah Cheang on Chinese material culture and British domesticity, and see p. 149 for further discussion of 'cosmopolitan domesticity').

Exhibitions such as the Colonial and Indian Exhibition in London in 1886 and the annual Ideal Home Exhibition, first held in London in 1908, were other important sites of spectacle and display that revealed the ways in which the home was an important site in fostering and articulating national and imperial power and identity. At the Colonial and Indian Exhibition, the popular song 'Home, Sweet Home' was performed at the opening ceremony, and the Prince of Wales was keen to stress to visitors that 'the Colonies ... are the legitimate and natural homes, in future, of the more adventurous and energetic portion of the population of these Islands' (British Parliamentary Papers 1887: xx). The Exhibition was widely reported throughout the empire, helping distant British subjects to imagine the links between their imperial homes and their home country (Blunt 1999). In India, for example, an article in the *Calcutta Review* reflected on the national and imperial importance of home in light of the opening ceremony:

> When we find on a great occasion that a picked elite of ten thousand of our countrymen and women are moved to tears at the sympathetic rendering by one woman's voice of the popular little song 'Home, Sweet Home,' we must feel convinced that both the sentiment and the music appealed to one of the strongest and most deep rooted of our national passions.
>
> (*Calcutta Review* 1886: 359; quoted in Blunt 1999)

From 1908, as Deborah Ryan explains, the Ideal Home Exhibition 'showed a more domestic view of Empire than the international exhibitions and frequently featured anthropological displays, representing "primitive" peoples not just as exotic others, but also as domesticated imperial subjects' (Ryan 1997a: 16; also see Ryan 1997b). Moreover, from 1926 to 1933, the Empire Marketing Board organized displays at the Ideal Home Exhibition 'that encouraged consumers to see the Empire as England's larder' (16). Metropolitan homes and domestic life were hence clearly influenced by the domestication of imperial subjects and by the rise of imperial consumption. The home was thus a material and an imaginative site invested with national and imperial significance, whereby domestic reproduction was intimately bound to power relations within, but also far beyond, the household.

Frontier home-making and American nation-building

From the sixteenth to the twentieth centuries, Western imperialism depended on trade in people, goods and ideas between the metropolis and the wider empire, as well as on resettlement and imperial home-making in places as diverse as plantation economies dependent on slavery in the West Indies and the American South; the frontier in the American West; white settler societies such as Australia, Canada, New Zealand, and South Africa; and long-term, but usually temporary, residence in places such as the Indian sub-continent. The colonization of the American West and the establishment of white settler colonies by modern European empires were, in turn, dependent on domesticated and gendered visions of home, nation, settlement and landscape. In his discussion of British settler society in the Cape Colony in the nineteenth century, for example, Alan Lester writes that 'women's provision of familiar domestic appearances, of household routine, and of reconciliatory diversions, lay at the heart of the settler community's reproduction' (Lester 2001: 73).

Colonization and settlement on the American frontier also evoked gendered and racialized images of home. Annette Kolodny, for example, writes that masculinist fantasies of conquest were underpinned by a pastoral yearning for 'land-as-woman' (Kolodny 1975). But, in contrast, women's fantasies of colonization and settlement were more domesticated, evoking memories of past homes alongside dreams of future homes. In written accounts about the frontier west, Kolodny argues that

> At the heart of their western vision was a fantasy of home that, though they did not acknowledge it, harked back to an earlier era. For, as a newly industrializing nation was fast eroding the economic functions of the home and consequently narrowing the scope of women's activity in general, the domestic novel of western relocation still suggested that the home, and, particularly, women's traditional role within it, held tangible significance.
>
> (Kolodny 1984: 165)

In her study of the writings by white Anglo emigrant housewives in North America – popularly known as 'pioneer women' – Janet Floyd explores the daily routines of domestic life and the ways in which they were tied to a wider project of nation-building based on territorial expansion, resettlement and the displacement of indigenous people (2002a; also see Floyd 2002b). Whereas the subjects of many studies of imperial domesticity concerned themselves with the employment and management of servants, such 'pioneer women' did domestic work themselves. Moreover, these women's autobiographical accounts of home-making enunciate 'the crucial importance of homemaking within the project of emigration and settling land, and a detail of the everyday round written to prove participation in that project' (2002a: 10).

As both Kolodny and Floyd show, 'The language of domesticity permeated representations of national expansion' (Kaplan 2002: 27). From the 1830s to the 1850s, 'the United States increased its national domain by seventy percent, engaged in a bloody campaign of Indian removal, fought its first prolonged foreign war, wrested the Spanish borderlands from Mexico, and annexed Texas, Oregon, California, and New Mexico' (26). Amy Kaplan studies the ways in which American national and imperial expansion was closely bound up with ideas about home and domesticity, by exploring 'how the ideology of separate spheres contributed to creating an American empire [and] how the concept of domesticity made the nation into home at a time when its geopolitical borders were expanding rapidly through violent confrontations with Mexicans and Native Americans' (26). Kaplan explores the mobile and shifting distinctions between the 'domestic' and the 'foreign' during this period of imperial nation-building, and points to the inherent contradictions of imperial domesticity. Through her analysis of the writings by Catharine Beecher and Sarah Josepha Hale, Kaplan argues that the discourse of domesticity performed a double, apparently paradoxical, movement that not only extended women's sphere of influence beyond the home and the nation, but simultaneously contracted it 'to that of policing domestic boundaries against the threat of foreignness' (28). For example, Sarah Josepha Hale, novelist and editor of the influential journal *Godey's Lady's Book*, launched a campaign in 1847 to make Thanksgiving Day a national holiday, which ran until it achieved its goal in 1863. Launched in the middle of the Mexican-American War, '[t]his effort typified the way in which Hale's map of woman's sphere overlaid national and domestic spaces; *Godey's* published detailed instructions and recipes for preparing the Thanksgiving feast, while it encouraged women readers to agitate for a nationwide holiday as a ritual of national expansion and unification' (Kaplan 2002: 35). For Hale, the Thanksgiving holiday was one important way to bolster the home and nation against the 'threat of foreignness' – a threat that she located both within as well as beyond the borders of the nation. For Hale, as Kaplan explains,

> the nation's boundaries not only defined its geographical limits but also set apart nonwhites within the national domain. In Hale's fiction of the 1850s, Thanksgiving polices the domestic sphere by making black people, whether free or enslaved, foreign to the domestic nation and homeless within America's expanding borders.
>
> (36)

The origins of Thanksgiving Day reveal the ways in which home on domestic and national scales were closely bound together, serving to domesticate particular stories about national origins and unity. The interior spaces, practices and family relationships of the domestic home thus helped to forge and to reproduce ideas about the nation as home. Moreover, such ideas about home were clearly racialized, privileging white American home life and national belonging whilst excluding black and Native Americans from the nation-as-home. But, whilst this example shows how white, bourgeois Americans sought to protect both the domestic and the national home from an internal and external 'threat of foreignness', by the end of the nineteenth century, the interior decoration of their homes often displayed a fascination with the 'foreign'. Kristin Hoganson (2002) explains that many bourgeois American homes in the period from 1865 to 1920 displayed a 'cosmopolitan domesticity', which was characterized by an 'enthusiasm for imported goods and styles perceived to be foreign, in large part because of their very foreignness' (57). As in many European countries at this time (see Research Box 7 by Sarah Cheang), bourgeois women would drape 'a packing box with gaudy fabric in hopes of making an Oriental "cozy corner"' (58), home decorating magazines heralded the interior styles of other countries, and merchants 'were quick to trumpet foreign provenance, seeing it as an enticement to purchase' (64). Cosmopolitan domesticity partly served to express the identities of their inhabitants, allowing them to distinguish themselves in terms of class. But cosmopolitan domesticity also, as Hoganson points out, reflected attempts to bring the world into the home: 'collecting and displaying imported objects provided a way to demonstrate a broad outlook, wide experience, and engagement with the world' (79). But such an engagement was limited. Cosmopolitan domesticity made gestures towards universalism even though it was based on unequal economic and political relations between countries, and was not only complicit with, but also contingent on empire. As Hoganson summarizes:

> It did not imply a belief in the essential quality of all human beings or a profound understanding of other nations and cultures. Nor did it necessarily imply a willingness to open the nation's borders to immigrants. The art experts who lamented the vitiation of 'authentic' styles outside of Europe promoted the idea that cosmopolitanism should be a testament to western knowledge, openness and modernity. Those who mixed and matched imported objects fabricated the exotic. Those who sought

imported items that had been crafted to suit their tastes or who arranged them so that they felt familiar domesticated the wider world, denying its difference and asserting their own appropriative power.

(80)

British homes in India

Unlike resettlement on the American frontier and in other settler societies, the homes of the British imperial elite in India were usually established on a long-term, but temporary, basis (see Gowans 2001, for more on the repatriation of British women after prolonged residence in India). British homes in India, and the domestic roles of British wives and mothers, were politically significant and contested, and have been studied by a wide range of geographers, historians and literary critics (see, for example, Sharpe 1993; George 1996; Grewal 1996; Blunt 1999; Procida 2002; Buettner 2004). Most imperial commentators agreed that the presence of British wives and mothers in India – women known as memsahibs – was necessary for establishing appropriately British homes and domestic life. For such commentators, the presence of British women as home-makers in India was essential not only for the reproduction of legitimate impe-rial rulers, but also for the reproduction of the domestic, social and moral values legitimating imperial rule. For example, the article in the *Calcutta Review* that reported the opening ceremony of the Colonial and Indian Exhibition in London in 1886 claimed that only the presence of English wives and home-makers in India could alleviate the homesickness of their husbands,

> among [whom] are hardworking, home-loving men – [whose] ideal of bliss is to consort with one to cheer them in health and nurse them in sickness, and who will tend their houses and administer their homes with discretion. All are Englishmen, and they love in their wives what is essen-tially English.
>
> (*Calcutta Review* 1886: 369, quoted in Blunt 1999)

But other imperial commentators claimed that British women and their homes in India led to separate spheres of exclusively British domestic and social life that provoked racial antagonisms between rulers and ruled, and ulti-mately contributed to the decline of the British Empire. Many commentators attributed the emergence of such separate spheres to memories of the 'Mutiny' of 1857–8.

As shown by Figure 4.2, the severity of the threat posed to British rule in India by the 'Mutiny' was most vividly depicted by the destruction of homes and by the fate of British women and children, particularly at Cawnpore (see Blunt 2000b). Middle-class British women who survived the five-month siege of Lucknow experienced the severity of the conflict most acutely on a domestic scale (Blunt 2000a). Six book-length diaries written by such women not only document everyday domestic life at a time of conflict but also represent the imperial conflict in domestic terms (Bartrum 1858; Case 1858; Harris 1858; Inglis 1892; Germon 1957; Brydon 1978). At the start of the siege, the diaries recorded some continuity of home life and imperial rule. But this changed dramatically after a few days when most Indian servants left their British employers. As Katherine Harris, the wife of an army chaplain, recorded in her diary soon after the start of the siege:

> People's servants seem to be deserting daily. We expect soon to be without attendants, and a good riddance it would be if this were a climate which admitted of one's doing without them; but if they all leave us, it will be difficult to know how we shall manage. Their impudence is beyond bounds: they are losing even the semblance of respect. I packed off my tailor yesterday: he came very late, and, on my remarking it, he gave me such an insolent answer and look, that I discharged him then and there; and he actually went off without waiting, or asking for his wages.
>
> (Harris 1858: 46–7)

For middle-class, married diarists such as Katherine Harris, imperial power was challenged most directly in a domestic sphere. For the first time, such women had to make tea, clean, wash their clothes and sometimes cook, although the wives of British soldiers were usually employed for this purpose. Most of the daily entries in the diaries record domestic hardships, domestic divisions of labour, and new routines of domestic work. As John Kaye wrote in his history of the uprising, 'our women were not dishonoured, save that they were made to feel their servitude' (Kaye 1876: 354).

In 1859, soon after the suppression of the 'Mutiny', a popular book of vignettes and illustrations by George Atkinson entitled *Curry and Rice' on Forty Plates; or, the Ingredients of Social Life at 'Our Station'* sought to reinscribe imperial and domestic power in the fictional upcountry station of Kabob (Atkinson 1859). Images of British women in their Indian homes were central to this endeavour. Most notably, as shown by Figure 4.3, images of British women with their servants embodied imperial power on a domestic scale and stood in stark contrast to the destruction of homes and the unaccustomed servitude for British women that had characterized many accounts of the 'Mutiny'. In many ways, the reconstruction of imperial power in India after 1858 was closely bound up with the reconstruction of imperial homes. Mary Procida describes the 'politicized imperial home' in British India, and writes that 'The most private and intimate spaces of the colonizers were themselves colonized by the demands of empire' (2002: 56; also see Stoler 2002, on imperial domesticity in the Dutch East Indies).

Sources as diverse as household guides for British women in India, memoirs, oral histories, literature and family photographs reveal the existence of 'empires in the home', and their management and regulation by women (see Box 4.1). Compounds were racially demarcated to house Indian servants and their families at a distance from the bungalows where British officials usually lived (and see Box 4.2 for more on bungalows). This reproduced on a domestic scale the racial distancing that underpinned colonial urbanism, whereby British cantonments and civil lines were located at a distance from the 'native' city. And yet such distancing was transcended on a daily basis as

Figure 4.3 'Mrs Turmeric, the Judge's Wife'. G. Atkinson (1859) *'Curry and Rice' on Forty Plates; or, the Ingredients of Social Life at 'Our Station'* (London: John B. Day).

Indian servants worked within British homes. In the most famous household guide about imperial domesticity, which was first published in 1888, Flora Annie Steel and Grace Gardiner wrote that a British home in India should represent:

> That unit of civilisation where father and children, master and servant, employer and employed, can learn their several duties. When all is said and done also, herein lies the natural outlet for most of the talent peculiar to women. ... We do not wish to advocate an unholy haughtiness; but an Indian household can no more be governed peacefully, without dignity and prestige, than an Indian empire.
>
> (Steel and Gardiner 1907: 7, 9)

By posing with their servants, photographs of British families in India served to reproduce an empire within as well as beyond the home, as shown by Figure 4.4.

Both in British India and elsewhere, imperial rule led to the development of new forms of settlement. Hill stations, for example, were a uniquely imperial

Figure 4.4 Colonel and Mrs Cotton at home in India, 1887. Photo 154/(58) Reproduced by permission of The British Library.

Box 4.1   EMPIRES IN THE HOME

This term refers to ways in which imperial power and authority were exercised on a domestic scale, particularly by women. It was a term adopted by a number of American writers in the nineteenth century (Kaplan 2002) and, more recently, by Rosemary Marangoly George (1996) in her study of imperial domesticity in British India. Sources including household guides, letters, diaries, memoirs and oral history interviews reveal that 'empires in the home' in British India revolved around the management of servants and anxieties about raising children. In the late nineteenth century, it was estimated that the smallest British household in India would require ten to twelve servants, while larger households might employ up to 30. Until the 1920s, it was still common to employ up to a dozen servants (Barr 1976; 1989). In comparable households in Britain at the same time, it would have been unusual to employ more than three to five servants. The high number of Indian servants was usually attributed to caste restrictions, whereby different individuals were employed to perform particular duties. With the exception of an ayah, who cared for babies and young children and also acted as lady's maid, the Indian servants employed by the British were male.

Indian servants can be seen as the domesticated outsiders of a British imperial imagination. Ann Laura Stoler writes that 'native' servants occupied a complex place within imperial homes:

> Represented as both devotional and devious, trustworthy and lascivious, native servants occupied and constituted a dangerous sexual terrain, a pivotal moral role ... it was their very domestication that placed the intimate workings of the bourgeois home in their knowing insurrectionary hands and in their pernicious control.
>
> (1995: 150)

As Stoler continues, 'native' servants were often treated as children:

> racialised Others invariably have been compared and equated with children, a representation that conveniently provided a moral

justification for imperial policies of tutelage, discipline and specific
paternalistic and maternalistic strategies of custodial control.

(1995: 150)

Household guides written for British women living in India not only
infantilized Indian servants, but also represented them as inferior to
their western counterparts (Blunt 1999). (Servants' own experiences
and opinions are usually 'unstoried' in the archives (Antoinette Burton,
2003); but see 'Memory-work in Java' by Ann Laura Stoler, with Karen
Strassler, in Stoler (2002), which analyses interviews with Javanese
servants who worked for Dutch employers in the Dutch East Indies.)

Imperial concerns about 'race', class and gender became particularly acute in the advice contained in household guides about raising
children in India. British children usually remained with their parents in
India for seven years, before being sent to school in Britain or in a hill
station for their health and education. Most household guides advo-
cated that babies should be breastfed by their mothers, but that wet-
nurses should be used in exceptional circumstances. Anxieties about
employing wet-nurses were long-established in Europe as well as
across the empire. In Europe, medical discourses focused on the risks
of a baby absorbing the 'personality traits' of a nurse, which, it was
thought, threatened to dilute aristocratic and bourgeois blood with
working-class breastmilk (Stoler 1995). In India these concerns were
articulated in terms of racialized as well as class differences and, by the
early twentieth century, Indian wet-nurses were rarely used (for a recent
film on the employment of an Anglo-Indian wet-nurse for a British baby
in Kerala in the 1950s, see the Merchant–Ivory production *Cotton
Mary*, 2002). Racialized anxieties about raising British children in India
continued beyond infancy, as shown by debates about whether Indian,
British or Anglo-Indian women should be employed as nannies or
ayahs. In their best-selling household guide, Steel and Gardiner wrote
that 'However good native servants may be, they have not the same upbringing and nice ways, knowledge, and trustworthiness of a well-
trained English nurse. Besides, native servants seldom have as much
authority over a child' (1907: 166–7). In similar terms, in a household
guide published in 1923, Kate Platt advised that 'children left to the

care and companionship of native servants run a serious risk of acquiring bad habits, of becoming unmannerly, and of developing in undesirable ways' (1923: 138–9), whilst anxieties about employing Anglo-Indian nannies revolved around British children picking up their distinctive accent (Blunt 2005).

Box 4.2   BUNGALOWS

In his classic study of the bungalow as a transnational domestic form, Anthony King explains that 'The bungalow, both in name and form, originated in India ... from the Hindi or Mahratti *Bangla* meaning "of or belonging to Bengal"' (King 1984: 14; also see King 1997 and Glover 2004). As Anne Campbell Wilson wrote in 1904, 'An Indian bungalow in the plains is a square, one-storied, flat-roofed house, with a pillared verandah at each side' (28), which was designed with many windows and doors to facilitate a through draught and with a wide verandah for shade. According to Maud Diver, 'an Indian bungalow is as exquisitely simple in construction as an English house is complex. It is not built to please the eye of man, but to shield his body from a merciless sun' (1909: 61).

Many British women tried to make their Indian bungalows as home-like as possible. Beatrix Scott, for example, had arrived in India in 1910 cherishing 'a suburban desire for a pretty little home whose carpets and curtains matched', and fondly remembered her bungalow:

> With the usual nostalgia the Briton has for his home surround-ings the living rooms had been furnished with as near an attempt to the English country home as could be accomplished. ... There were chairs upholstered in rose strewn chintz, soft grey carpets blushing with rosebuds, creamy curtains appliqued with big loose-leafed roses, and when later I unpacked my own pictures and household treasures I quite liked my home.
> (Scott Papers, Centre of South Asian Studies, University of Cambridge)

Writing in the 1970s, describing her life in India fifty years earlier, Mrs Parry also described her bungalow as homelike:

> Our existence in the bungalow resembled life at home ... It was a charming bungalow most comfortably arranged. When we were alone in the evening after dinner, if the picture of the drawing room could have been televised ... it would have disclosed a prosaic vision of two people sitting by the fire reading in a room which might have been in a country house anywhere in England. ... On the walls there were various prints of favourite 'Old Masters' which added to the homelike appearance, and of course there were large vases of flowers. The spaniels sat at our feet gazing into the fire with typical Cocker Spaniel expression. Even the pressure lamps did not give our position on the map away, as their staring white light was veiled by circular pink silk shades.
>
> (Memoir by Kapi, Parry Papers, Centre of South Asian Studies, University of Cambridge)

In contrast, Drusilla Harrington Hawes believed that 'we would have been wiser to furnish and live like Indians, instead of making frantic efforts to create English homes out of Indian bungalows ... But this was unthinkable in the days of the Raj' (Harrington Hawes Papers, Mss Eur C533, Oriental and India Office Library Collections, British Library).

The domestic form of the bungalow travelled from India to the West from the 1860s, and came to be closely associated with the growth of suburbia. As King explains, 'The suburb was instrumental in producing the architectural form of the bungalow, just as the bungalow was instrumental in producing the spatial form of the suburb' (1997: 56). The bungalow was socially and architecturally significant for two main reasons. First, it represented a spatial change that was both vertical and horizontal, moving from 'living in houses of three or four storeys in large, densely populated cities and towns, dependent on the railway, to ones of two or, increasingly one storey, in more spacious outer suburbs dependent primarily on the car' (56). Second, the bungalow represented a social and architectural change, 'from living in an architectural form containing many

households – such as (in Britain) the ubiquitous Georgian or Victorian terrace, or perhaps semi-detached villa, or (in North America)
the apartment, row house or "brownstone" – to living in the modern
detached or semi-detached house or bungalow containing only one'
(56).

King also studies the cultural significance of the bungalow today,
as shown by his account of bungalows in multicultural Australia. He
explains that Australia is one of the most 'suburbanized' countries in
the world, and that suburban bungalows reflect changing patterns of
migration and resettlement: 'Like earlier Greek, Italian, Turkish,
Vietnamese and other immigrants before them, Chinese families
move into the suburbs, cranking up the levels of difference and
diversity in the already multicultural landscape. ... Vestiges of Anglo-
Tudor are erased and the Anglo-Indian-Californian-Australian multicultural bungalow is step by step transformed into an ordinary
Chinese "house"' (1997: 77; also see Jacobs (2004) on representations of house and home for a Chinese family in Australia in the film
Floating Life, and Mitchell (2004) on migration from Hong Kong to
Vancouver in the 1990s and the emergence of 'monster houses' in
elite suburbs, as discussed in Chapter 5).

form of settlement, and were thought to represent a home away from home.
From the 1820s, the British established approximately sixty-five urban settle-
ments at elevations from four to eight thousand feet above sea level in India
(Kennedy 1996; hill stations were also established throughout South and
South East Asia in, for example, the Dutch East Indies, the Philippines, and
Japan). Simla, in the lower Himalayas, was the oldest, largest and best-known
hill station, and was the summer capital of British India from 1864 to 1939.
Hill stations were represented as the most suitable location for British settlement, largely because the 'hills' were seen to provide a healthier environment
than the 'plains', particularly for women and children (Kenny 1995; Blunt
1999). Moreover, the houses, social life, climate and landscape associated
with hill stations were imagined by British residents to be more home-like
than the rest of India (see Duncan and Lambert 2003, for more on home and
the aesthetics of imperial landscapes). Unlike 'classical' bungalows elsewhere
in India, European-style houses, often built in a tudor or gothic-revival style,

were rented for the season. Their names, such as 'Moss Grange', 'Ivy Glen', and 'Sunny Bank', evoked memories and imaginations of Britain as home, as did the transportation of material objects to make these houses as home-like as possible. Amongst many other things, Steel and Gardiner advised British women that 'Carpets for the sitting-rooms and all curtains must be taken, piano, small tables, comfortable chairs, nicknacks, ornaments ... chair backs, tablecovers, something to cover the mantelpiece, and possibly a few pictures' (199). They suggested that a 'lady', three or four children, and an English nurse would require eleven camel-loads of luggage to set up temporary home in the hills.

Domestic life both in the metropolis and across the empire was intimately shaped by imperial politics, and imperial home-making was a critical site for both domestic and imperial reproduction. And yet, imperial homes were also sites of contestation, particularly in relation to the domestic roles of women far from home and their domestication – alongside children and servants – as imperial subjects. Research on imperial domesticity in a wide range of disciplines and contexts reveals the gendered and racialized interplay of home, nation and empire and its contested embodiment, particularly by wives and mothers.

HOMELAND, NATION AND NATIONALIST POLITICS

Homeland, nation and homely belonging

Imaginative and material geographies of home have been important in nationalist as well as imperial politics, often through representations of the nation as homeland. As Thembisa Waetjen writes, 'The idea of a homeland is of a place embodying social essences, cultural or historical, that legitimate claims to a natural sovereignty. A homeland is the landscape also of historical memory that offers tangible images of rootedness and grounded community' (1999: 654). The idea of a homeland is thus bound up with the politics of place, identity, and collective memory (see Chapter 5 for more on home and homeland over transnational space). Alongside the evocation of home and landscape, claims to 'natural sovereignty' are often closely tied to claims to *national* sovereignty, as the idea of a homeland – whether remembered from the past, existing in the present, or yet to be created – is mapped onto national space. In diverse contexts, the spaces of home, homeland and nation inscribe gendered and racialized geographies of inclusion and exclusion.

One of the most significant examples of the national (and nationalist) resonance of the idea of a homeland is represented by the German term *Heimat*.
'Historically,' writes David Morley, 'the ideas of home and homeland have
perhaps been most emphatically intertwined' in this concept (2000: 32).
Existing alongside ideas of a German fatherland, *Heimat* invokes 'notions of an
idealized mother and an idealized feminine' (Blickle 2002: ix), and was used
by the National Socialist Party in the 1930s and 1940s to signify an Aryan
sovereignty that was based on racial and ethnic exclusivity. But the potent idea
of *Heimat* existed long before the rise of fascism, and has undergone a process
of political and cultural rehabilitation since the late 1950s. In his research on
artistic depictions of *Heimat*, Christopher Wickham writes that the term is no
longer necessarily bound to an idea of the nation, but rather invokes longing
and belonging and serves 'as a point (or set of points) of reference for individual social identity' (1999: 10; elsewhere, the state has established homelands. See Box 4.3 on homelands in South Africa, and see p. 169 for discussion
of homeland security in the United States since 9/11).

Box 4.3   HOMELANDS IN SOUTH AFRICA

In some places, the state itself has created new homelands. Most infamously, the apartheid state in South Africa established homelands as
part of its policy of racial and ethnic segregation. As Waetjen writes,
'Territorial re-tribalization of Africans in *bantustans* ("homelands")
gained under apartheid a new and ominous significance, manifested
most concretely in the progressive vision and determination to create
independent states for Africans based on ethnic membership' (1999:
657). Although the apartheid state envisioned an exclusive, white Afrikaner society, it was dependent on cheap black labour. As Waetjen
continues, 'Relocating and reorganizing the African population into
ethnic homelands involved limiting mobility, integration and resistance while yet serving the needs of capital through a migrant labour
system'. Following the Urban Areas Acts (1945) and the Group Areas
Acts (1951), the Bantu Authorities Acts in the 1950s organized tribal
governments in each *bantustan*. The creation of *bantustans* had two
main effects: first, from 1960 to 1983, mass displacement as 3.5
million South Africans were relocated; and, second, 'the solidification

of a bantustan political elite whose authority was explicitly vested in ethnic nationhood' (Waetjen 1999: 658). Today, in post-apartheid South Africa, the Freedom Front aims to establish a new, independent homeland, but this time for Afrikaners. Developing the small Afrikaner settlement, Oriana, 'the aim is to create, in miniature, an Afrikaans haven which the brave, or foolhardy, believe might one day grow into an Afrikaans homeland' (Hope 2003). A similar vision of an exclusively white and reactionary rural homeland is the attempt by the far right in Britain to establish an Aryan community in Essex (Ryan 2003). The ideas about colonization, settlement and self-help that underpin attempts to establish white homelands at Oriana and near Chelmsford echo the earlier, and much larger-scale, examples of white colonization and settlement (clearly, not all rural settlements and colonies were politically reactionary, as shown by anarchist settlements such as Whiteway Colony set up in Britain in the nineteenth century. See Blunt and Wills 2000 for further discussion).

Discourses about the nation as homeland are often characterized by the gendered use of domestic and familial imagery. As Anne McClintock explains,

> Nations are frequently figured through the iconography of familial and domestic space. The term 'nation' derives from 'natio': to be born. We speak of nations as 'motherlands' and 'fatherlands.' Foreigners 'adopt' countries that are not their native homes, and are 'naturalized' into the national family. We talk of the Family of Nations, of 'homelands' and 'native' lands.
>
> (1993: 63; quoted in Cowen 2004: 762; also see Walter 1995; Radcliffe 1996)

Rather than view such familial and domestic images of the nation as benign, Deborah Cowen writes that they enable, legitimize and naturalize 'the production and reproduction of domination and exploitation' (2004: 762). As a result, 'the familial imaginary [is] one of the most loaded and hidden ideologies, which has a tremendous effect on the configuration of social and economic life' (762).

Exploring such familial imagery, and the ways in which it is closely bound to ideas of home, Ghassan Hage describes 'homely belonging' as the most common of nationalist discourses, whereby national subjects identify with a homeland that is a 'bountiful and fulfilling ... secure, pleasing, and gratifying space'. Whilst images of the nation as fatherland correspond to the ordered and empowered spaces of governmental and sovereign belonging, the site of homely belonging is most commonly figured through images of the nation as motherland. According to Hage, 'all the qualities that are valued in the homeland are those that are normally (that is, within patriarchal discourse) associated with mothering: protection, warmth, emotional and nutritional security' (Hage 1996: 473). The figure of the mother is symbolically central to national identity and nationalist discourse, and national subjects are positioned in relation to the maternal nation as children, particularly as sons. The motherland is often embodied as a maternal subject such as Mother India, Mother Ireland or Mother Russia (see, for example, Lyons 1996; Thapar-Björkert and Ryan 2002). Imagining a national homeland as the motherland is closely tied to imaginings of nature and landscape as female and is embodied by iconic figures of both real and imagined women.

Home and anti-imperial nationalism

Gendered as female and embodying the nation as home, Bharat Mata – Mother India – was one of the central symbols of anti-imperial nationalism in India and has been an important force in the rise of Hindu nationalism since the 1980s (Corbridge 1999). But in anti-imperial nationalist politics in India and elsewhere, the home was politically important in material as well as symbolic terms. Exploring the imaginative and material contours of home for middle-class Bengalis, for example, Dipesh Chakrabarty and Partha Chatterjee show the importance of domestic space and social relations in forging nationalist politics. Chatterjee argues that middle-class Bengali homes, the place of women within them, and anxieties about the westernization of women and domesticity were all vitally important in shaping the spiritual domain of anti-imperial nationalism. As he writes, 'it was the home that became the principal site of the struggle through which the hegemonic construct of the new nationalist patriarchy had to be normalized' (1993: 133; also see Walsh 2004 on household guides written for Indian women in the late nineteenth century). And yet Chatterjee's influential work arguably 'reproduces the domestication of women' (Legg 2003: 12) by consigning them to the home.

A wide range of other research has explored women's agency in anti-imperial nationalist politics, often within as well as beyond the home. Albeit with very different objectives and characteristics, nationalist homes – like imperial homes in the empire and the metropolis – were politically significant. As Suruchi Thapar-Björkert writes, 'Not only were the public / private boundaries blurred, the domestic arena became an important site for the steady politicisation of women's consciousness' (1997: 494). New constructions of femininity and motherhood helped in this process of politicization, and the home itself became an important site in the nationalist struggle. In his research on the importance of the home in anti-imperial nationalism in Delhi from 1930 to 1947, Stephen Legg highlights 'the ways in which women achieved agency in a nationalist movement that, while encouraging female participation, attempted to spatially delimit this activity to the home' (7). As Legg shows, 'women helped to politicize the home and assert their agency in a space often read as one of silence and subjection' (23) by, for example, organizing their homes as unofficial political headquarters, and wearing home-spun cloth, or khadi, which was part of Gandhi's campaign to replace cheap European imports with Indian-made goods.

The home is also an important site of memory in nationalist politics, as shown by Antoinette Burton's (2003) examination of the writings by three Indian women – Janaki Majumdar, Cornelia Sorabji and Attia Hosain – either in or about India in the 1930s. In different forms, the writings by these three women articulated memories of home to claim a place in history that was situated

> at the intersection of the private and the public, the personal and the political, the national and the postcolonial. All three were preoccupied with domestic architecture, its symbolic meanings and its material realities, because they were keenly aware that house and home were central to their social identities and the cultural forms through which they experienced both family life and national belonging.
>
> (4)

For example, Burton traces the 'domestic genealogies' that run throughout the memoir written in 1935 by Janaki Majumdar, the daughter of a prominent Indian nationalist. By describing the dwelling-places of her family over time and in different locations, Majumdar's 'Family History' is 'an essay in remembrance whose objects of imagination and desire are the houses of her family's past' (32). As Burton shows, this memoir interweaves a family history with a national history through its attempt 'to furnish a domestic genealogy of Indian nationalism' (62). The home was also an important site for imperial nationalist as well as anti-imperial nationalist politics, as shown by the Anglo-Indian attempt to set up an independent homeland and nation in the 1930s (Box 4.4).

Box 4.4   ANGLO-INDIAN HOME-MAKING AT MCCLUSKIEGANJ

Anglo-Indians represent one of the oldest and largest communities of mixed descent in the world (see Box 4.7 for more on home, identity and mixed descent, and Research Box 7 by Akile Ahmet on young men, mixed descent and home) and, before Independence, they often identified a British fatherland and an Indian motherland as home (Blunt 2002; for more on how public political discourses shifted from identifying with the fatherland to the motherland, and how this was both reproduced and resisted within Anglo-Indian homes, see Blunt 2005). Some Anglo-Indians who did not feel at home in India established a homeland called McCluskieganj in Bihar from 1933, whereas many more migrated after Independence (Blunt 2003c, 2005). Settlement at McCluskieganj was promoted in terms of a nostalgic desire for home that was rooted in both Britain and India. In 1934, an article in the monthly journal the *Colonization Observer* described the aims of the settlement scheme:

> our Community being the only Homeless one in this vast sub-Continent are, firstly, colonizing with the express object of establishing a Home for itself, secondly, of securing a definite stake in our own Country thereby becoming Indians proper, without losing our identity as Anglo-Indians, and lastly, by getting together we automatically open up fresh avenues of employment for the future generations.
>
> (*Colonization Observer*, May 1934: 1)

McCluskieganj represented a dream for Anglo-Indian independence that was located within British India and remained loyal to the British Empire. And yet, the vision of Anglo-Indian home- and nation-building at McCluskieganj also appealed to an Indian desire for home. As the founder of the scheme, Ernest Timothy McCluskie, wrote in 1935,

> Every Indian, whatever his station in life, can proudly say he has a piece of land and a hut, which he calls by the sweet word 'Home' ... but, alas, we who are bred and born in this country cannot say we have a home. This, therefore, is the real foundation of our scheme. *'To help you to have a Home,'* and to feel the joy and pride of possession of a *real home* of your very own.
>
> (CSI 1935)

This desire for home was encapsulated by the Hindi word *mooluk*, which suggested a place of origin, belonging and authentic identity, and located 'home, sweet home' for Anglo-Indians within the Indian motherland (see Figures 4.5 and 4.6). Alongside a desire to

Figure 4.5 McCluskieganj as 'Home, Sweet Home', 1939. *Colonization Observer*, March–April 1939. Reproduced courtesy of Alfred de Rozario.

44   THE COLONIZATION OBSERVER   MARCH—APRIL 1939

Figure 4.6 McCluskieganj as *Mooluk*, 1939. *Colonization Observer*, March–April 1939. Reproduced courtesy of Alfred de Rozario.

establish an Anglo-Indian *mooluk*, settlement at McCluskieganj was legitimated through appeals to an imperial paternal heritage, likened to white colonization in places such as Australia, New Zealand and Canada, and used to imagine a future for India as a dominion within the British Empire. Images of Anglo-Indian men as hardy pioneers, striving to colonize part of India just as their European forefathers had colonized the empire, underpinned the dream for independence at McCluskieganj. But while a collective memory of European paternal descent was embodied by Anglo-Indian men, the existence of an Indian maternal ancestor was usually erased. Instead of invoking an Indian ancestor, India itself was described as the motherland and the natural environment was described in maternal terms. Moreover, representations of Anglo-Indian women as pioneering home-makers meant positioning them within a collective memory of European colonization, echoing depictions of pioneer women in other colonized places. By sharing in the agricultural labour and home-building of their husbands, fathers and brothers, Anglo-Indian women were crucially important in promoting McCluskieganj as a home, homeland and nation that remained loyal to the British Empire.

Home, nation and homeland (in)security

Home and homeland are powerful spatial imaginaries that also articulate contemporary ideas about the nation. As we discuss in Box 4.5, Billig's ideas about 'banal nationalism' have been interpreted in relation to the home through a focus on comfort and domestic material culture. In more overt and less 'banal' ways, however, ideas about the nation as home and homeland have been scripted in terms of security, as graphically shown by the politics of homeland security in the United States since 9/11.

Box 4.5 HOME, COMFORT AND 'BANAL NATIONALISM'

In his research on 'prized possessions' belonging to members of working- and lower-middle-class households in the central western suburbs of greater Sydney, Greg Noble traces the ways in which 'homes articulate domestic spaces to national experience' (2002: 54). Figure 4.7 shows the ways in which national symbols can serve not only to make domestic spaces home-like, but also to domesticate ideas of the nation. Noble argues that

a language of comfort, of being 'at ease' with and through these objects, permeated the narratives that people offered and

Figure 4.7 'Griffith Marsupial, Frank and Pierina Bastianon' by Gerrit Fokkema, 1987. Reproduced by permission of State Library of New South Wales and Gerrit Fokkema.

seemed to provoke the idea that in making themselves 'at home' in a specific, domestic space, these people also seemed to be making themselves 'at home' in a larger social space.

(54–5)

Rather than interview people about their active national affiliation and their sense of national identity, Noble is interested in the ways in which a 'banal nationalism' becomes naturalized through domestic material cultures as well as everyday life, beliefs and habits (for more on 'banal nationalism', see Billig 1995). The homes of Noble's interviewees 'were filled with objects with a clear "Australian" symbolic content – landscape paintings, ornaments and trinkets, and so on, which evoked images of Australian places and history' (55), but the resonance of such objects with ideas about the nation as home remained largely submerged or 'backgrounded' within the home. Noble argues that such a process of 'backgrounding' underpins a sense of comfort at home both on domestic and national scales:

'The extent to which we feel a sense of being "at home" ... rests on the capacity of objects to withdraw, to become 'invisible' elements of an embodied, practical knowledge of familiar space' (58). In contrast to the ways in which the politics of homeland security – and Figures 4.1 and 4.2 – *fore*ground the relationships between home and nation, everyday images and objects also serve to naturalize such relationships, but do so in less overt ways. As Noble continues, 'This backgrounding of objects in the space of the home is fundamental not only to the naturalizing of the home as an inhabited space and the family as a collectivity, but is also crucial to an understanding of our everyday experience of the nation' (58; Noble also writes that many migrants from non-English-speaking countries and their children 'often had a different relationship to an Australian national identity than that experienced by those of Anglo-Australian ancestry' (61). See Chapter 5 for more on domestic material cultures within diasporic homes.)

According to Erin Manning, the intimate and exclusionary connections between modern, normalizing discourses of the home and the nation-state are underpinned by the desire for security, which is 'manifested as a collective fear and a resentment of difference – fear of that which is not us, not certain, not predictable. The quest for protection against the unknown results in a tightening of the borders of the nation, the home, and the self' (2003: 33). In this section we explore the ways in which the politics of homeland security in the United States today are bound up with material and imaginative geographies of home. We argue that the ways in which 'the borders of the nation, the home, and the self' are mapped onto each other serve to bolster the exclusionary and contested distinction between the 'domestic' and the 'foreign'.

The term 'homeland security' has been widely used in the United States since 9/11. The Department of Homeland Security (www.dhs.gov) is responsible for protecting the nation against terrorist and other threats. Many academic centres have been established to help in this endeavour (see, for example, the webpages of the National Academic Consortium for Homeland Security at http://homelandsecurity.osu.edu/NACHS, and the Homeland Security Institute at http://homelandsecurityinstitute.org), and GIS techniques have been widely marketed and employed as important tools for protecting the homeland (see, for example, www.esri.com). Like 'Ground Zero', the term 'homeland' has entered the everyday lexicon (Kaplan 2003). But despite its wide currency, little attention has been paid to the term itself. When President Bush used the term in a speech soon after 9/11, Amy Kaplan writes that

> it struck a jarring note as an unfamiliar way of referring to the American nation ... Why not *domestic security*? *Civil defense*? *National security*? How many Americans, even at moments of fervent nationalism, think of America as a homeland? How many think of America as their country, nation, home, but think of places elsewhere as their historical, ethnic, or spiritual homeland?
>
> (2003: 85)

'Homeland security' in the United States represents, for William Walters, a clear example of 'domopolitics'. According to Walters, 'Whereas political economy is descended from the will to govern the state as a household, domopolitics aspires to govern the state like a home' (2004: 237), and involves refiguring the relationships between citizenship, state and territory in order to rationalize 'a series of security measures in the name of a particular conception of home' (241). Describing the nation-state as a 'homeland' mobilizes 'powerful affinities with family, intimacy, place':

> the home as hearth, a refuge or a sanctuary in a heartless world; the home as *our* place, where we belong naturally, and where, by definition, others do not; international order as a space of homes – every people should have (at least) one; home as a place we must protect. We may invite guests into our home, but they come at our invitation; they don't stay indefinitely. Others are, by definition, uninvited. Illegal migrants and bogus refugees should be returned to 'their homes.' Home is a place to be secured because its contents (our property) are valuable and envied by others. Home as a safe, reassuring place, a place of intimacy, together-ness and even unity, trust and familiarity.
>
> (241; see Chapter 5 for more on domopolitics and immigration policy)

In other words, domopolitics – as shown by the politics of homeland security in the United States since 9/11 – is both legitimized by, and also enshrines, normative assumptions about the home requiring protection to ensure its familiarity, safety and security.

Although the term 'homeland' suggests historical ties to a particular place, it has only recently been widely used in the United States. The 'home front' during the Second World War, as in other places, characterized the American nation as separate and safe from more distant battlefields. During the Cold War, '[t]he domestic response to nuclear threat ... was called "civil defense," not homeland defense' (85). In addition to these examples, Kaplan shows that the term 'homeland' – with its connotations 'of native origins, of birthplace and birthright' and its appeals to 'common bloodlines, ancient ancestry, and notions of racial and ethnic homogeneity' (86) – stands in stark contrast to 'traditional images of American nationhood as boundless and mobile' (86). Examples of such traditional images of American nationhood – including 'a nation of immigrants,' 'a melting pot,' the 'western frontier,' 'manifest destiny,' and 'a classless society' – 'all involve metaphors of spatial mobility rather than the spatial fixedness and rootedness that homeland implies' (86). Why, Kaplan asks, has the term 'homeland' become so widespread and significant since 9/11, particularly given its lack of historical resonance in the United States? Recognizing the ways in which the nation as home depends upon a notion of the 'foreign', Kaplan poses further important questions: 'in reimagining America as the homeland, what conceptions of the foreign are implicitly evoked? What is the opposite of homeland? Foreign lands? Exile? Diaspora? Terrorism?' (86). What is at stake in identifying the nation as homeland?

Kaplan argues that the recent articulation of the American nation as 'the homeland' is intimately tied to the 'mobilization and expansion of state power', whereby

A relation exists between securing the homeland against the encroachment of foreign terrorists and enforcing national power abroad. The homeland may contract borders around a fixed space of nation and nativity, but it simultaneously also expands the capacity of the United States to move unilaterally across the borders of other nations.

(87)

In similar terms, Walters also writes that domopolitics invokes the meaning of 'domo' 'as conquest, taming, subduing; a will to domesticate the forces which threaten the sanctity of home' (2004: 242):

Domopolitics is not reducible to the Fortress impulse of building walls, strengthening the locks, updating the alarm system. It contains within itself this second tendency which takes it outwards, beyond the home, beyond even its own 'backyard' and quite often into its neighbours' homes, ghettos, jungles, *bases*, slums. Once domopolitics extends its reach, once it begins to take the region or even the globe as its strategic field of intervention, then the homeland becomes the home front, one amongst many sites in a multifaceted struggle.

(242)

The idea of the nation as homeland, in other words, has material implications both within and beyond its borders. Not only has the idea of the homeland been mobilized to legitimize the 'war on terror' beyond the United States, but it has also led to the strengthening of state power over people within the United States itself. 'At a time when the Patriot Act has attacked and abrogated the rights of so-called aliens and immigrants,' Kaplan writes, '[and] when the U.S. government can detain and deport them in the name of homeland security, the notion of the homeland itself contributes to making the life of immigrants terribly insecure' (87).

Such homeland insecurities were also plainly evident in the rise of racial violence in the United States after 9/11. At least five people were killed, almost 1,000 separate 'bias incidents' were reported in the two months after 9/11, and many more unreported cases of 'racial shame, uncertain immigration status, and the inaccessibility of law enforcement resources to many communities of color make it certain that the actual number of bias incidents is far higher' (Ahmad 2002: 103–4). Muneer Ahmad points to the gendered nature of this hate violence, with Muslim women in the United States, and elsewhere, reporting that their head scarves had been violently torn off. As Ahmad continues, '[f]or many Muslim women, the only means of protecting against such physical violence was to stay at home. ... [I]n the same moment that we decry the Taliban's cruel restrictions on the mobility of Afghan women, our racial oppression confines women in the United States to their homes as well. We have engaged in our own form of purdah' (110). Ahmad contrasts the veil and the American flag as two 'overdetermined symbols' since 9/11, and argues that the widespread 'embrace of the flag' by many Arabs, Muslims and South Asians represents 'a forced reveiling of the community' (110). Like the display of patriotic symbols such as 'God Bless America' spelt out in lights in the front yard (Figure 4.8) and yellow ribbons to remember American soldiers posted

Figure 4.8 'God Bless America', Oklahoma, November 2001. Photograph by Steve Liss, Time Life Pictures, Getty Images. Reproduced by permission of Time Life Pictures / Getty Images.

overseas, the prominent place of the American flag in front of many homes binds the domestic and familial sphere to the imagined space of the nation as homeland. Ahmad argues that this 'reveiling' is not the only choice facing people who face racist oppression, and points instead to the importance of forging coalitions across different communities, as shown by, for example, the Hate Free Zone Campaign in Seattle, and the 'Circle of Peace' established by a multi-racial group around a mosque in Chicago (112).

As Ahmad and Kaplan argue, homeland security is thus inseparably bound up with homeland insecurity. For Kaplan, the very idea of the homeland evokes 'prior and future losses, invasions, abandonment' and is profoundly uncanny, 'haunted by all the unfamiliar yet strangely familiar foreign specters that threaten to turn it into its opposite' (89). As she continues, 'Although homeland security may strive to cordon off the nation as a domestic space from external foreign threats, it is actually about breaking down the boundaries between inside and outside, about seeing the homeland in a state of constant emergency from threats within and without' (90). The home itself, as well as the homeland, is thus 'in a continual state of emergency', and 'vast new intrusions of government, military, and intelligence forces' (90) are mobilized and legitimized to protect both home and homeland from these threats. As Deborah Cowen puts it, '[h]ome is mobilized in a terrifying slippery scalar slope that moves from its private-residential meaning to a sinister national longing, as in "homeland defense"' (2004: 756). Different scales of home and homeland collapse into each other, whereby:

> [An] obsession with membership, security, and order inevitably seems to collapse the imagined homeland with the literal home itself, defining similar objectives for each site or scale. Indeed, these ideological articulations of national belonging produce conceptions of 'domestic life' that simultaneously define public cum national and private cum familial meanings. To be an American citizen is to know that the home and the homeland both require defence; both require clear boundaries and strict surveillance; both require order and hierarchy; both require unity and purification.
>
> (757)

As a contemporary echo of the Thanksgiving Day campaign by Godey's Lady's Book in the nineteenth century, which related the domestic to the national home in the pages of a women's magazine, the American Ladies' Home Journal assessed homeland security in its 2002 rating of cities (Kettl 2003). Both terrorist and other threats reveal the ways in which the politics of homeland security in the United States are predicated upon, and also perpetuate, a politics of insecurity. For many people, the homeland remains a profoundly unhomely place.

INDIGENEITY, HOME AND BELONGING

As Kaplan observes, there is a 'terrible irony' for Native Americans in the current use of the term 'homeland' to describe the United States (2003: 87). In the final part of this chapter, we consider the violence of dispossession that resulted from imperial resettlement and nation-building, and the unhomely homes to which many Native Americans and other indigenous people were, and are, forcibly relocated. As Porteous and Smith explain,

> Whether in Canada, Australia, or in the United States in the nineteenth century, or in the tropical forests of South America, Africa, and Southeast Asia today, colonization and economic exploitation result in geopiratic domicide – large-scale removals, disempowered victims, loss of identity, and often a common good rhetoric as justification.
>
> (2001: 85; see Box 4.6 on domicide)

Histories and memories of dispossession and forced relocation have important implications for social justice, belonging, and the politics of home today. We explore the ways in which past and present geographies of dispossession unsettle ideas about home in relation to the city aand the nation. We focus in particular on Australia and Canada to consider the long-term effects of settler colonialism and nation-building on indigeneity, home and belonging.

Box 4.6 DOMICIDE

This term was coined by Douglas Porteous and Sandra Smith (2001) and refers to 'the deliberate destruction of home by human agency in pursuit of specified goals, which causes suffering to the victims' (12; also see Read, 1996, on lost homes). Ranging from a single dwelling to an ethnic homeland (such as Kosovo and East Timor in 1999), homes are destroyed by a variety of means: 'from warfare through economic development and urban renewal to the creation of roads, airports, dams, and national parks. Too frequently, the elimination of home or homeland is justified as being in the public interest or for the common good' (ix). Porteous and Smith estimate that at least thirty million people across the world are victims of domicide (ix). They distinguish between 'extreme' and 'everyday' domicide. The former refers to 'major, planned operations that occur rather sporadically in time but often affect large areas and change the lives of considerable numbers of people' (64), such as war and the forced resettlement of indigenous people. In contrast, the latter 'occurs continuously all over the world and can affect everyone except the wealthy and those who are its perpetrators. Unlike extreme domicide, the everyday variety comes about because of the normal, mundane operations of the world's political economy', and results from 'urban redevelopment, economic restructuring, or placement of public facilities' (107).

Victims of domicide, according to Porteous and Smith, are not the same as refugees: first, most domicide victims remain within their own country, whereas refugees cross international borders (see Chapter 5); and, second, most refugees flee from warfare or environmental degradation, but only the former of these may involve domicide. Unlike the widespread destruction of home through disasters such as the Asian Tsunami in December 2004, Hurricane Katrina in the United States in 2005, and the South Asian earthquake in 2005, domicide results directly from the human agency of powerful people involved in corporate, political or bureaucratic projects. Although the effects of both domicide and 'natural' disasters reflect deep-rooted social injustice, Porteous and Smith argue that injustice is a 'causal issue' in the former but not the latter case (16).

The importance of the home as a meaningful place – 'perhaps the most meaningful of all' (7) – is central to the notion of domicide. Drawing in particular on humanistic traditions of writing about home, Porteous and Smith contrast two meanings of home: first, an outward-looking focus on 'home as centre' – 'a place of refuge, freedom, possession, shelter and security'; and, second, an inward-looking focus on 'home as identity', bound up with 'family, friends and community, attachment, rootedness, memory, and nostalgia' (61). They argue that the spatial, symbolic and psychosocial meanings of home mean that domicide may result in 'the destruction of a place of attachment and refuge; loss of security and ownership; restrictions on freedom; partial loss of identity; and a radical de-centring from place, family, and community' (63).

International organizations such as the Centre on Housing Rights and Evictions (COHRE) work 'to promote and protect the right to housing for everyone, everywhere' (www.cohre.org). Although COHRE does not use the term 'domicide', its work seeks to prevent what Porteous and Smith describe in their book. COHRE explains that 'The right to housing is one of the most widely violated human rights. The United Nations estimates that around 100 million people worldwide are without a place to live. Over one billion people are inadequately housed' (www.cohre.org). People who are forcibly evicted from their homes are victims of domicide. COHRE defines forced evictions as 'the removal of people from their homes or lands against their will, directly or indirectly attributable to the State' (www.cohre.org), a practice that is widespread in both developed and developing countries (see Table 4.1 for the common reasons given by the state to legitimize forced evictions). 'Under international human rights law,' COHRE explains, 'you have the right to be protected against forced evictions', which is part of the broader right to housing (www.cohre.org). As listed in Table 4.2, many other rights are also violated through the practice of forced eviction.

Table 4.1 Common reasons for the state to legitimize forced evictions

Development and infrastructure projects (e.g. dam construction)

Prestigious international events (e.g. Olympics)

Urban redevelopment or city beautification projects

Conflict over land rights

The removal or reduction of housing subsidies for low-income groups

Forced population transfers and forced relocations in the context of armed conflict

Separation of ethnic or racial groups

Refugee movements

Reclaiming public land

Source: www.cohre.org

Table 4.2 Human rights violated by forced evictions

The right to housing

The right to non-interference with privacy, family and home

The right to the peaceful enjoyment of possessions – to limit the potential for organized resistance, many forced evictions occur without warning, forcing people to abandon their homes, lands and worldly possessions

The right to respect for the home

The right to freedom of movement and to choose one's residence

The right to education – often children cannot attend school due to relocation

The right to life – violence during the forced eviction which results in death is a common occurrence, with developers going to many lengths to obtain land

The right to security of the person – rarely are displaced groups provided with adequate homes or any form of compensation, which renders them vulnerable to homelessness and further acts of violence

Source: www.cohre.org

Dispossession

As Ceridwen Spark explains, 'indigeneity is constituted by the notion that one is already at home' (1999: 58). Modern European imperial expansion led to the dispossession of indigenous people from their land in Australasia and the Americas. This dispossession often resulted from a fiction of 'terra nullius' (land unoccupied), whereby indigenous occupation and ownership was unrecognized in the face of settler repossession and the imposition of colonial sovereignty and property laws. According to Aileen Moreton-Robinson,

> Australia was a multicultural society long before migrants arrived. It is estimated that over 500 language groups held title to land prior to colonization. Indigenous people owned, lived on, were taught to know and belonged to particular tracts of 'country' which is the term used to refer to one's territory / land of origin or a person connected to the same piece of land. Indigenous people's sense of belonging is derived from an ontological relationship to country derived from the Dreaming, which provides the precedents for what is believed to have occurred in the beginning in the original form of social living created by ancestral beings.
> (Moreton-Robinson 2003: 31. Also see Moreton-Robinson 2000)

Both in Australia and elsewhere, the fiction of 'terra nullius' 'was modified to embrace not only the occupancy of land but its *appropriate* use' (Johnson 1994: 147). In British Columbia, for example, Cole Harris explains that the emergence of a 'reserve geography' in the nineteenth century

> was the product of the pervasive settler assumptions, backed by the colonial state, that most of the land they encountered in British Columbia was waste, waiting to be put to productive use: or, where Native people obviously were using the land, that their uses were inefficient and therefore should be replaced. … Such assumptions, coupled with self-interest and a huge imbalance of power, were sufficient to dispossess Native people of most of their land.
> (2002: 265; also see Blomley 2004c)

By the 1930s, more than 1,500 small reserves came to represent 'native space' in British Columbia, with profound and long-term implications for the lives and livelihoods of indigenous people. As Harris writes,

The reserve was now the principal locus of Native life, occupied for much more of the year – in many cases for the entire year – than the winter village had ever been. By the beginning of the twentieth century, most people lived in detached log or frame houses that were intended for nuclear families and for year-round occupancy. From the reserves they still moved out to local resource procurement sites, but to far fewer such sites than formerly because so many of them had become private properties where Native people were trespassers.

(2002: 288)

The dispossession of indigenous people from their land and livelihoods in countries such as Australia and Canada continues to pose crucial questions about home and belonging. As Moreton-Robinson writes, 'Who calls Australia home is inextricably connected to who has possession, and possession is jealously guarded by white Australians' (2003: 27). In a Canadian context, Harris argues that '[Native people] are here in a more rooted sense than any of the rest of us, their former lands are the basis on which the whole country has been built, and the case, in justice, for redressing the drastic imbalances of colonialism is overwhelming' (2002: 303).

Unhomely homes

Settler colonialism not only led to the dispossession of indigenous people from their land, but also led to the forced relocation of many people from their homes. Most infamously,

In 1838, the United States Army rounded up the Georgia Cherokee and kept them for several months in disease-ridden camps. Then they forced them to trek, at bayonet point, throughout the whole winter across 1500 km of rough territory to Oklahoma. One-quarter of the 15,000 Cherokees died along this Trail of Tears.

(Porteous and Smith 2001: 78–9)

Although 'there was much less contempt, less overt cruelty, and few long-distance trails of tears in Canada' (81), other forced relocations took place. In British Columbia, for example, many indigenous children were sent to residential schools, where they were 'subjected to rigorous space-time disciplines, watched, weaned as much as possible from their Nativeness, and remade, at least in principle, as English-speaking members of a civilized, modern society' (Harris 2002: 269). In Australia,

> Legislation and state policies served to exclude Indigenous people from participation as citizens through their removal to reserves, missions and cattle stations where their everyday lives were lived under regimes of surveillance. ... Other Indigenous people were stolen from their families and placed in institutions or adopted by white families.
>
> (Moreton-Robinson 2003: 33)

For the first sixty years of the twentieth century, the 'stolen generation' of children of part-Aboriginal and part-white descent were forcibly removed from their families to be raised in residential homes, mission schools and white families (see Box 4.7 on home, identity and mixed descent, and Research Box 7 by Akile Ahmet). In New South Wales alone, more than 10,000 Aboriginal children were removed from their homes and families. Moreover, an estimated one in six Aboriginal children were removed from their parents in the twentieth century, compared to one in three hundred non-Aboriginal children (MacDonald 1995). In the Northern Territories, the Kahlin 'Half-Caste' Home in Darwin and The Bungalow in Alice Springs both opened in 1913 in an attempt to assimilate part-Aboriginal children to white society by removing them from their homes and families and by seeking to sever ties with Aboriginal culture and history. Most children were removed from their families by local policemen, acting in the role of Aboriginal Protector. Some children tried to escape and return home (see Box 4.8 on *Rabbit-Proof Fence*), but many never saw their families again (MacDonald 1995).

Box 4.7   HOME, IDENTITY AND MIXED DESCENT

The relationships between home and identity are a recurrent theme in work on, and by, people of mixed descent (also see Research Box 7 by Akile Ahmet on young men, mixed descent and home). Alongside a wide literature on 'inter-racial' partnering, parenting, fostering and adoption, there is a growing literature on home and identity that extends beyond family relationships to explore a wider sense of home and belonging. According to Joanne Arnott, 'possibly the most difficult issue for people of mixed heritage is that of belonging': of finding a place to call home (1994: 266). In a book entitled *Scattered Belongings*, Jayne Ifekwunigwe writes that 'In the de/territorialized places, which "mixed race" cartographers map, the idea of "home" has, by definition, multilayered, multitextual and contradictory meanings' (1999: xiv–xv). Such complex and multiple mappings of home often reflect a sense of identity and belonging that is both personal and transnational, as shown by a range of autobiographical writings, particularly by women. For example, Velina Hasu Houston writes that 'As an Amerasian who is native Japanese, Blackfoot Indian, and African American, I am without the luxury of state ("home"). ... Home is sanctuary from the world, but it is not found in one physical place or in a particular community' (1996: 276, 278).

Drawing on interviews with self-identified women of 'mixed race' in Toronto, Minelle Mahtani argues that the term is used in different ways as a 'linguistic home' that can create 'new geographies of inclusion' (2002: 487). Mahtani critiques popular discourses that are characterized by 'a relentless negativity' in their portrayal of 'mixed race' individuals as out of place or with no place to call home (470). In similar terms, Jill Olumide writes that 'one of the salient features of the social construction of mixed race has been its characterisation as a marginal, detached and confused state in which individuals so designated are condemned to wander in search of belonging and acceptance' (2002: 5). Other research has focused on the material as well as metaphorical geographies of home and identity for people of mixed descent. Alison Blunt, for example, explores the spatial politics of home for Anglo-Indians in the fifty years before and after Independence, and shows how political debates about home, nation and empire were both reproduced and recast on a domestic scale (Blunt 2005). The materialities and social relations of everyday domestic life had a wider political significance for this community of mixed descent, as shown by the distinctiveness of Anglo-Indian cuisine, western eating practices and household structure, and the mobilization of ideas about home in debates on national identity that articulated a dual loyalty to Britain as fatherland and India as motherland.

Box 4.8   *RABBIT-PROOF FENCE*

In 1996, the Aboriginal author Doris Pilkington Garimara published a novel entitled *Follow the Rabbit-Proof Fence*, which was made into a film called *Rabbit-Proof Fence* in 2002. The novel and the film tell the story of Garimara's mother (Molly Craig), her younger sister and their cousin, who, like other children of the Stolen Generation, were forcibly removed from their family in 1931 in Jigalong, Western Australia. Aged 14, 11 and 8, the three girls were taken from their mothers to be trained as domestic servants at the Moore River native settlement north of Perth. But they escaped and walked home for nine weeks and almost 1,500 kilometres through the desert, with the rabbit-proof fence as their only guide. As Garimara explains, the fence represented the 'most solid link to home': 'They grew up beside the fence, it meant love, warmth, home. If you are ever lost then once you have found the fence it will take you all the way home' (quoted in Barkham, 2002). Molly was retaken nearly ten years after her escape, and returned to the Moore River settlement with her two daughters. Again she escaped to Jigalong with her younger daughter, but four-year old Doris remained at Moore River and was not reunited with her mother for twenty years.

The film focused attention on the pain and suffering of members of the Stolen Generation like Doris and Molly. But the film also led to a conservative backlash, with some commentators defending the government policy. One reviewer notes that the film not only represented the dispossession of Aboriginal people, but that it also had powerful resonances with the injustices suffered by asylum-seekers in Australia today: 'Images from Rabbit-Proof Fence of children in detention are especially poignant, as the United Nations, the Roman Catholic Church, charities, and international human rights groups have been protesting to the government over its treatment of asylum-seekers, which includes locking up children in a desert camp' (Barkham, 2002).

When Kahlin 'Half-Caste' Home closed in 1939 and The Bungalow closed in 1942, most part-Aboriginal children in the Northern Territories were sent to mission homes or to be raised by white families. A 1957 Melbourne newspaper article, for example, described the adoption of three Aboriginal girls from the Northern Territories by the white Deutsher family: 'Mr Deutsher said last night that the way to solve the native problem was to bring them into the homes of white people so that they could be thoroughly acclimatised' (reproduced in MacDonald 1995). As Rowena MacDonald explains, 'Some of the children were taught to be so ashamed of their Aboriginal descent that they pretended all their lives that they were really white or of some other nationality. Some children never even knew that they were Aboriginal. They would never know that they had other names, other relatives and other homes, in places they had never heard of' (MacDonald 1995: 59). Since the early 1980s, the indigenous organization Link-Up has worked

> to assist these people to find their way home to their natural families, their Aboriginal communities and culture. ... Link-Up assists in working through the effects of removal and separation to prepare people for a successful reunion, and helps them to come home by accompanying them as they meet their families for the first time.
>
> (Kendall 1995: 72)

From 1998 to 2003, Link-Up facilitated 887 family reunions and homecomings (Department of Foreign Affairs and Trade, Australian Government: www.dfat.gov.au/facts/separated_children.html).

In 1997, the Australian Human Rights and Equal Opportunity Commission published *Bringing them Home*, which is the report of the two-year National Inquiry into the Separation of Aboriginal and Torres Strait Islander Children from their Families (Commonwealth of Australia 1997; the report is available on the Reconciliation and Social Justice Library website at: www.austlii.edu.au/au/special/rsjproject/rsjlibrary/hreoc/stolen). The oral testimony quoted in the report vividly conveys the pain of separation, the overcrowded and harsh living conditions in most institutional homes, and the sexual abuse often suffered within such homes and while living with white families (the National Library of Australia began a 'Bringing them Home' oral history project in 1999. Unlike the confidential testimonies collected by the Inquiry, this oral history project aims to create an open record. Visit www.nla.gov.au for further information, and also see Bird 1998). According to the Report,

> Because the objective was to absorb the children into white society, Aboriginality was not positively affirmed. Many children experienced

contempt and denigration of their Aboriginality and that of their parents or denial of their Aboriginality. In line with the common objective, many children were told either that their families had rejected them or that their families were dead. Most often family members were unable to keep in touch with the child. This cut the child off from his or her roots and meant that the child was at the mercy of institution staff or foster parents. Many were exploited and abused. Few who gave evidence to the Inquiry had been happy and secure. Those few had become closely attached to institution staff or found loving and supportive adoptive families.

(Bringing them home: the effects)

The effects of the forced removal and separation of the 'stolen generation' from their families were far-reaching and long-term:

The Inquiry was told that the effects damage the children who were forcibly removed, their parents and siblings and their communities. Subsequent generations continue to suffer the effects of parents and grandparents having been forcibly removed, institutionalised, denied contact with their Aboriginality and in some cases traumatised and abused.

(Bringing them home: the effects)

The Report recommended that 'those organizations directly implicated in the forced removals [should] deliver appropriately worded apologies to indigenous Australians', and that there should be a 'National Sorry Day', 'which would offer both the Stolen Generation and other Australians a chance to remember the pain inflicted on Aborigines and Torres Strait Islanders by past policies' (Gooder and Jacobs 2002: 207). Within a year after the Report was published, a number of state leaders, police forces, and church groups had offered apologies and, in 1998, the first National Sorry Day was held and the first Sorry Book was opened for settler Australians to sign. Although the Prime Minister, John Howard, expressed his regret and personal sorrow, he did not apologize because the Australian Government 'does not support a formal apology to Indigenous people. Such an apology could imply that present generations are responsible and accountable for the actions of earlier generations even though those actions were sanctioned by the laws of the time and were believed to be in the best interests of the children' (Department of Foreign Affairs and Trade, Australian Government: www.dfat.gov.au/

facts/separated_children.html). As Haydie Gooder and Jane Jacobs (2002) argue, the contested politics of the apology in a reconciling nation are intimately tied to wider politics of belonging and non-belonging.

The politics of home and belonging

Both the dispossession of indigenous people from their land and livelihoods and the forced relocation of indigenous children from their homes and families continue to underpin an unsettled politics of home and belonging in countries such as Australia and Canada today. In contrast to an ideology of assimilation, the politics of home and belonging are often articulated in terms of reconciliation, social justice, and cohabitation. In this section, we consider the politics of home and belonging for settlers and indigenous people as they are materialized in relation to the city and the nation.

In his research on urban land and the politics of property in Vancouver, Nicholas Blomley (2004c) unsettles the city, in part through his analysis of the dispossession and displacement of indigenous people. Whereas dispossession 'refers to the specific processes through which settlers came to acquire title to land historically held by aboriginal people' (109), displacement 'refers to the conceptual removal of aboriginal people from the city, and the concomitant "emplacement" of white settlers' (109). Like other cities in Canada and elsewhere, Vancouver was 'superimposed upon a network of ancient native villages, resource sites, and symbolic landscapes' (110). Endemic poverty and the lack of land and resources on reserves has meant that, by 2001, 'half of all Aboriginal people in Canada were urban, one-quarter of them living in ten urban areas' (113). Blomley explores the imposition of colonial ownership through private property law, and the materialization of this in the urban – and often the domestic – built environment. As he writes,

> Colonial cities, put simply, cannot be conceived as native spaces because they have so obviously been occupied, built upon and 'improved.' In her exploration of European conquests in the Americas, Patricia Seed explores the culturally diverse ways in which the possession of the New World was justified and rendered natural. English claims to territory, she shows, were enacted through apparently 'mundane activity,' where the placement of objects – houses, fences, gardens – signified ownership: 'Englishmen occupying the New World initially inscribed their

possession ... by affixing their own powerful cultural symbols of ownership – houses and fences – upon the landscape'.

(119; Blomley is quoting Seed 1995: 19, 25)

Both in Vancouver and elsewhere, as Blomley shows, 'Native people have not
only resisted historic dispossessions, they have also ... remapped a continuing
native presence, reminding observers that the settler-city not only *was*, but still
is, native land' (131; also see Johnson 1994; Jacobs 1996; and our discussion
of The Block in Sydney in Box 2.7).

   Such contested politics of land, ownership, occupation and belonging also
unsettle ideas about the nation as home. In their analysis of sacredness and
identity in Australia, for example, Ken Gelder and Jane Jacobs (1998) employ
Freud's ideas about the uncanny to unsettle both the home and nation. As they
explain, the value of this notion 'is that it refuses the usual binary structure
upon which much commentary on Aboriginal and non-Aboriginal relations is
based. We often speak of Australia as a 'settler' nation, but the 'uncanny' can
remind us that a condition of unsettledness folds into this taken-for-granted
mode of occupation' (24). As part of their analysis of 'the troubled entanglements of possession and dispossession, settlement and unsettlement' (36),
Gelder and Jacobs explore a number of Australian ghost stories (also see
Research Box 8 by Caron Lipman). Margot Nash's film *Vacant Possession* (1996),
for example, revolves around the return of its non-Aboriginal protagonist,
Tessa, to her childhood home in Botany Bay. As Gelder and Jacobs explain,
'The film focuses on this now dilapidated and abandoned family home and
the traumas that unfold when she returns there after her mother's death' (36).
Tessa is haunted by the memory of falling pregnant to an Aboriginal boy
when she was young, and her father's violent and racist response that led him
to shoot the boy and drive Tessa from the family home. The climax of the film
involves Tessa, her father and an Aboriginal neighbour called Millie sharing a
meal on her return:

   The traumatic histories of nation and family are drawn together around
   the table, and the consequences are nothing less than Gothic: a tempes-
   tuous storm erupts to shake the very foundations of the family home in
   Botany Bay. The unlikely trio ... takes shelter in the cellar, during which
   time father and daughter, Aboriginal and non-Aboriginal neighbour,
   confront and lay to rest the ghosts of their past. The house is blown away

in the tempest, as if dispossession must be shared equally in order for
Tessa to achieve a 'proper' homecoming.

(36–7)

The film is about reconciliation:

it speaks quite self-consciously to the national condition, using Tessa's
homestead as an image of Australia itself. ... [R]econciliation, as it is
conceived in this film, rests on a non-Aboriginal character returning
home in order to become *both immersed and dispossessed*: to become
homely and homeless at the same time. ... That is, this non-Aboriginal
character is reconciled by becoming 'Aboriginal' in a postcolonial sense:
immersed in the landscape, but dispossessed of property: all in the frame
of Botany Bay (un)settlement.

(37)

In his book on belonging in Australia, Peter Read writes that 'I want to feel I
belong here while respecting Aboriginality, neither appropriating it nor being
absorbed by it' (2000: 15), and asks 'How can we non-Indigenous Australians
justify our continuous presence and our love for this country while the Indigenous people remain dispossessed and their history unacknowledged?' (1).
Drawing on personal accounts from a wide range of non-Aboriginal Australians, Read refers to 'the house of Australia', within which every room 'is not
just occupied, it is shared' (2). David Crouch (2004) explores the resonances
between Read's account and literary depictions of dwelling in Australia, particularly Tim Winton's novel *Cloudstreet* (1998). As Crouch explains, 'there is the
anxious question of dwelling comfortably, in settled repose, in sanctuary, in
homes that are set in nature, and upon a land, which was stolen before it was
settled' (49). Crouch shows that the difficulties of dwelling are often articulated
through 'ecological tropes of the ground, the land and the natural world':

How can we speak of dwelling in the terms of private sanctuary spaces, or
the ground, or the lie of the land, when the foundations of this dwelling
are literally 'grounded' in a landscape which is originally not ours? When
it is the true and sacred, untranslated, dwelling place of Indigenous
people who hail the natural world as both their kin and culture? How does
this awareness then affect one's sense of an interior's felicitous space?

(49–50)

CONCLUSIONS

In this chapter we have begun to mobilize the home beyond the scale of the house, charting the material and imaginative geographies of home in relation to imperial power, nationalist resistance, the nation as homeland, and the politics of indigeneity. One theme that has run throughout the chapter is the way in which the materialities and imaginaries of home are closely connected rather than distinct, as shown by ideas about establishing and maintaining 'empires in the home', by the politicization of the domestic sphere in anti-imperial nationalist politics, and by the contemporary implications of 'homeland security'. In each case, we have argued that there is a double movement between the domestic home and the nation and/or empire beyond: not only have the wider spatial imaginaries of nation and empire been reproduced and recast within the domestic sphere, but the material and imaginative geographies of home and family have also been central in underpinning and articulating the wider nation and/or empire. In other words, home-spaces and home-making practices are intimately bound together over a range of scales, and are closely shaped by the exercise of power and resistance and by what is imagined as 'foreign' or unhomely.

This chapter has thus explored the ways in which the home is an intensely political site. We have sought to unsettle the idea of home as a fixed and stable location by exploring its inclusions, exclusions and contestations. Imperial resettlement and nation-building, for example, often led to the dispossession of indigenous people, many of whom were forced to relocate to a range of unhomely homes. At the same time, contemporary concerns about 'homeland security' in countries such as the United States have scripted various racialized inclusions and exclusions. In both material and imaginative terms, the home is an important site for articulating wider debates about national belonging today. Building on our discussion in Chapter 3, we have explored the ways in which the politics of home and belonging are gendered, racialized and underpinned by the assumed heterosexuality of family life on national and imperial as well as domestic scales. This chapter has also introduced the importance of home on a *transnational* scale, as shown by imperial resettlement and home-making, and by the politics of indigeneity in relation to such imperial nation-building. Chapter 5 turns to focus on the unprecedented scale of migration and resettlement in the contemporary world, investigating the implications of transnational mobility for home and identity, home-making practices, and the links between home and homeland.

Research box 6   DRAGONS IN THE DRAWING ROOM: RESEARCHING GENDER AND CHINESE MATERIAL CULTURE IN THE HOME

Sarah Cheang

From tea drinking to porcelain collecting, Chinese products have a history of fashionability in Britain through which paradigms of gender identity have been played out. In my doctoral work, I examined how Chinese material culture was used to articulate British femininities in the late nineteenth and early twentieth centuries. What did it mean for a woman to arrange flowers in a Chinese vase, sew Chinese embroideries onto satin curtains, or lounge in Chinese pyjamas on Chinese cushions? Or, to be more precise, how were feminine involvements with Chinese things understood, and what did these potentially exotic encounters say about Britishness and gender construction?

That these encounters were largely domestic in nature soon became an extremely important factor in my work. If domesticity is one of the defining elements of feminine identity (Davidoff and Hall 2002), then the home as a location for Chinese things took on a number of key roles within the interconnected ideologies of gender, class and nation. Methodologically also, domestic settings became a useful way to focus my study. Asking which spaces in the home were particularly associated with women, and examining the incidence of Chineseness in those spaces, enabled a surprisingly wide range of investigations, from the picturing of Pekingese dogs in the drawing room to the 'Chinese' sitting room of Queen Mary's Doll's House.

A domestic interior can be a primary location for expressions of identity through the design and distribution of things (Csikszentmihalyi and Rochberg-Halton 1981). My own research had previously focused on collecting as a reification of self, an expression of deep-seated cultural and psychological desires and an extension of personal boundaries from the body into the room itself (Stewart 1993; Cheang 2001). The making of a home can also be understood in this way, as a profound and multi-layered projection of self. The interdisciplinary nature of my study – embracing art, craft, fashion, furniture and interior design – further stressed the indivisibility of clothing, interior design and identity, creating 'home' as a fascinating arena for encounters with Chineseness and British femininities. Representations of women owning and using Chinese things in interior design advice literature and women's magazines indicated that a powerful array of modern, feminine subject identities could be assumed through the construction of 'Chinese' interiors. Furthermore, fictional works such as the *Forsyte Saga* (Galsworthy 1924) presented 'Chinese' drawing rooms as moral barometers and devices that gave a character depth and substance through imagined acts of home-making. Thus, 'home' featured strongly in fictional and non-fictional processes of self-fashioning that knowingly used the idea of home-as-self as a theatre for home-as-deliberate-projection-of-self.

Fashions for Chinese things were also connected with colonial war and political volatility. Armed conflict in China brought an influx of traded goods and loot, and economic instability in China made embroidered goods more available to western consumers. The meanings of Chinese things were therefore closely tied to their contexts of acquisition, so that the picturing of Chinese textiles, ceramics and carpets in late nineteenth- and early twentieth-century British homes underscored an understanding of 'home' as formed under the particular historical conditions of British imperialism. This fluidity of movement between empire and home, between the colony and the domestic, meant that a 'Chinese' drawing room could also be viewed as a form of female imperial agency. It seemed that, in principle at least, men could go East, whilst women stayed at home creating 'Chinese' rooms, revealing the home as an important site for feminine access to empire. My conclusions therefore led me back to the curious doubling of the word 'domestic' as meaning 'nation' and 'feminine'. This constant movement between outside and inside, exterior and interior, home and self, was a constant reminder that British feminine identities were tied to homes constructed in relation to colonial encounter.

Sarah Cheang completed her DPhil in the History of Art at the University of Sussex in 2003. Her thesis was entitled 'The Ownership and Collection of Chinese Material Culture by Women in Britain, c.1890–c.1935'. She is now Lecturer in Cultural Studies at London College of Fashion. Her recent and current research centres on two projects, both of which enable her to explore the boundaries of the self through fashion and material culture. The first examines the use of Chinese textiles in European and American domestic interiors between 1860 and 1950, where Chinese embroideries were ambiguously viewed as garments, as soft furnishings, as art and as bric-a-brac. The second

Figure 4.9   'Miss Ella Casella with Two of Her Chow Chows', *Ladies' Field*, 23 July 1898: 260.

Research box 7   **HOME AND IDENTITY FOR YOUNG MEN OF MIXED DESCENT**

Akile Ahmet

In a recent piece of research on the educational heritage of mixed-heritage school pupils, Tickly *et al.* (2004) found that 'white/black pupils face specific barriers to achievement. Low expectations of pupils by teachers seem often based on stereotypical views of the fragmented home backgrounds and confused identities of white/Black Caribbean children' (2). In response to this research, and in an attempt to challenge such stereotypical views, I am investigating the intersections of home, family and identity for young men of mixed descent in London today.

My research is situated within wider debates about the home and family, embodied masculinities, and mixed descent, identity and performativity. Whilst a wide range of research has studied young men in public spaces, and whilst there is a growing literature on home and masculinity (including Tosh 1999; Varley and Blasco 2001), there have been few studies of young men at home. At the same time, research on mixed descent has explored identity, education and parenting, often focusing on particular groups of people such as women, parents and children, and particular spaces such as the home and the school (including Tizard and Phoenix 2002; Ali 2003; Blunt 2005). Whilst there have been important autobiographical and auto-ethnographic work on mixed descent, home and identity (including Ifekwunigwe 1999; Mahtani 2002), most of this has focused on the lives and experiences of women. My research explores the ways in which embodied identities of mixed descent and the geographies of home can be understood in terms of performativity. I am interested in the performativity of 'race', ethnicity and nationality alongside gender and sexuality, and I am studying the material as well as the affective spaces of home.

My research investigates the role that mixed descent plays in shaping the everyday experiences of young men at home. I am interested in the ways in which young men regard the home in relation to other places and spheres of life, their relationships with their families, and the ways in which the material cultures of home might manifest or erase their mixed descent. I plan to interview young men of mixed descent who attend Tower Hamlets Summer University, where I am working in the summer teaching a course on 'race' and ethnicity. I will also ask the young men to complete written or audio diaries. I will also interview officers and members of two organizations for people of mixed descent: People in Harmony and Intermix. As the website of Intermix makes clear, home and family are important concerns for people of mixed descent and their parents:

> There is a need for mixed race individuals to learn as much as possible about themselves and how they want to be perceived by society. There is a need for parents of mixed race children to understand the diverse cultural needs of their children and to try and find as much information as possible to help them provide their children with a balanced cultural upbringing. There is a need to try and educate racially mixed couples about the potential obstacles they may encounter when raising a mixed race child.
>
> (www.intermix.org.uk)

I am particularly interested in the ways in which young men of mixed descent perceive a 'balanced cultural upbringing' and the ways in which this might be reflected, resisted or suppressed within the home.

Akile Ahmet is a PhD candidate in the Department of Geography, Queen Mary, University of London. Her thesis is entitled 'Home and Identity for Young Men of Mixed Descent', and involves interviews and autobiographical research with young men about home as a material, emotional and performative space.

Research box 8 THE DOMESTIC UNCANNY: CO-HABITING WITH GHOSTS

Caron Lipman

My research aims to examine the affective, non-rational and intangible aspects of people's relationship to domestic space through a qualitative investigation of the suburban haunted home in contemporary Britain. Its primary focus is to discover the ways in which people who believe their homes to be haunted negotiate the experience of co-habiting with ghosts, and what insights such an experience reveals about people's embodied, emotional, spatial and temporal connections and interchanges with home as both a physical place and a wider social myth or ideal.

I have always had a fascination for haunted homes, not least because of the apparent ubiquity of such anomalous experiences recounted by a wide cross-section of the population, experiences which are often spontaneous and shared by others, which are at odds with mainstream and scientific conceptions of knowledge, and which cannot always be bracketed off as counter-culture phenomena such as New Age beliefs.

My thesis develops out of a particular moment within cultural and human geography that attends to the invisible or unseen aspects of human relationships to place, bringing together interest in issues of affect, memory and subjectivity, scrutiny of the binary between nature and culture, attempts to re-materialize geography, and a move towards new forms of phenomenology with their engagement with performative and mobile, everyday processes.

The haunted home has enjoyed a long-standing position as a motif within society, crossing a span of narratives, from anecdotal local stories shared informally between family and friendship networks, to the established gothic traditions of literature and film. Despite this, there has been no major anthropological or sociological study into people's experiences of living in haunted homes. Cultural geographers who have recently expressed an interest in ghosts and hauntings have tended to focus upon public metropolitan spaces and to employ the ghost as a metaphor rather than an intangible presence in its own right. In contrast, this project will contribute to the growing literature on the material and immaterial geographies of the home. I have chosen to describe experiences within suburban homes because these are, after all, where most people live; and, despite a growing body of work showing the complex etymology of suburban living, suburbs are still commonly viewed as rather bland, homogeneous places.

A key challenge will be to test out and reflect upon methods for researching the intangible aspects of people's experience of place. Given the issue is not to prove or disprove the existence of supernatural phenomena, how do we situate the ghost, or give it a kind of agency? How do we account for the ghost in the relationship between the home and the householders?

At this early stage in the development of this doctorate, I have a number of other, related, research questions in mind. The issue of how to re/present immaterial geographies connects to a wider exploration of the techniques used to record the presence of ghosts and responses to being haunted; this, in turn, fits with work on the historical and developing relationship between technology, place and supernatural belief.

The project will also ask: how does a haunting affect existing beliefs and attitudes towards people's relationship to their home? How far does an uncanny presence provide a magnifying focus on existing relationships between embodied/emotional experiences of home, and recurring themes such as ownership, co-habitation and privacy? How far are such relationships changed through the experience of being haunted? How far are people's experiences of haunting influenced by or described as part of wider myths, narratives and beliefs? What is the relationship between the biographies of houses, current occupiers' sense of their home's history, and their own personal biographies? And what role does personal and social memory play in haunting?

I hope to move towards some answers — or at least a more informed set of questions yet to be imagined — through the course of my ethnographic project, applying a mix of creative methodologies to a number of in-depth case studies.

Caron Lipman is currently a PhD candidate in the Department of Geography, Queen Mary, University of London. The title of her thesis is: 'The Domestic Uncanny: Co-habiting with Ghosts.' She also works as a journalist, specializing in urban policy and the environment.

5

TRANSNATIONAL HOMES

This chapter is about the effects of transnational mobility on feeling at home or not at home, at home in more than one place, or homeless. Whilst we introduced transnational geographies of home in the previous chapter in relation to imperial resettlement and home-making, we now turn to the unprecedented scale of migration and resettlement in the contemporary world. Thus the chapter considers home from the perspective of people who migrate across national borders: those who leave home for primarily economic reasons, those who are forced to do so because of war, persecution or dispossession, and the effects of migration on those who remain at home. We return to some of the themes that we have discussed in previous chapters, including domestic architecture and domestic work, the material and symbolic intersections of home and homeland, and the profound and long-term implications of dispossession and displacement. Building on our earlier discussion, we consider the ways in which material and imaginative geographies of home, and the lived experiences of home on a domestic scale, are mobilized, reproduced and recast through transnational migration and resettlement as well as transnational circuits of capital and ideas. We argue that transnational homes are shaped by the interplay of both mobile and located homes and identities and by the processes and practices of home-making both within particular places and across transnational space.

The multi-scalarity of home is particularly apparent in relation to transnational homes, as shown by the high-rise and the bungalow as transnational domestic forms (see Chapters 3 and 4), the transnational employment of domestic workers, and the transnational resettlement of exiles, refugees and asylum seekers within new homes, camps and detention centres. We are particularly interested not only in the ways in which diasporic, transnational and global imaginaries influence, and are themselves influenced by, everyday, domestic experiences and practices, but also in wider questions about the very idea of home, of what home might represent, and where home might be located. Some of these questions are listed in Box 5.1

Box 5.1   HOME AND TRANSNATIONAL MIGRATION: QUESTIONS IN THE LITERATURE

The wide and growing literature on transnational migration, transnational communities, and diaspora raises important and challenging questions about home, including:

Brah (1996)

- 'When does a place of residence become "home"?' (1)
- 'Where is home?' (192)
- 'When does a location *become* home? What is the difference between "feeling at home" and staking claim to a place as one's own?' (193)

Ahmed (2000)

- 'What does it mean to be at home? How does it affect home and being-at-home when one leaves home?' (77)

Al-Ali and Koser (2002):

- 'How do transnational social fields and practices manifest themselves in daily lives, and how (if at all) do they impact on abstract conceptualizations of home?' (7)
- 'Does the existence of [transnational] communities necessitate a reconceptualization of the notion "home"? To what extent is "home" for transnational migrants no longer tied to a specific

---

geographical place? To what extent do transnational migrants conceive of more than one "home", with competing allegiances changing through time?' (8)

Ahmed et al. (2003a)

- 'What ... is the relationship between leaving home and the imagining of home? How are homes made in the context of migration? And what, having left home, might it mean to return?' (8)

Fouron (2003)

- 'Is the conceptualization of "home" as a fluid concept unbounded by the barriers of national sovereignty capable of birthing a new "internationalist" movement in the twenty-first century?' (209)

As shown by Box 5.1, research on home and transnational migration raises important questions that destabilize a sense of home as a stable origin and unsettle the fixity and singularity of a place called home. Moreover, such questions also suggest that ideas of home are relational across space and time, are often shaped by memories of past homes as well as dreams of future homes, and bring together both material and imaginative geographies of residence and belonging, departure and return. Transnational homes are thus shaped by ideas and experiences of location and dislocation, place and displacement, as people migrate for a variety of reasons and feel both at home and not at home in a wide range of circumstances.

The chapter has four parts. We begin by exploring the transnational relationships between home and homeland, home-making practices across diasporic space, and the politics of home for different generations living in diaspora. We then turn to the personal and political significance of home for exiles, asylum seekers and refugees and investigate contested forms of residence and home-making in refugee camps and through policies of dispersal and repatriation. In the third part of the chapter, we focus on transnational geographies of home for women who migrate as domestic workers. We end by exploring what it might mean to feel at home in a global city or suburb.

HOME, HOMELAND AND TRANSNATIONAL MIGRATION

According to Nadje Al-Ali and Khalid Koser, 'The changing relationship between migrants and their "homes" is held to be an almost quintessential characteristic of transnational migration' (2002: 1). The lived experiences and spatial imaginaries of transnational migrants revolve around home in a range of ways: through, for example, the relationships between home and homeland, the existence of multiple homes, diverse home-making practices, and the intersections of home, memory, identity and belonging. In this section we explore the symbolic attachments to a homeland that are materialized through diasporic home-making and return journeys, as well as the economic and political connections across multiple homes that characterize diverse transnational communities (see, for example, Rapport and Dawson 1998; Al-Ali and Koser 2002; Ahmed et al. 2003b; Yeoh et al. 2003).

The term 'diaspora' refers to a scattering of people over space and transnational connections between people and places. According to Bronwen Walter, 'Diaspora involves feeling "at home" in the area of settlement while retaining significant identification outside it' (2001: 206). The lives of transnational migrants are often interpreted in terms of 'roots' and 'routes', which articulate two ways of thinking about home, homeland and diaspora (Clifford 1997; Gilroy 1993). Whilst 'roots' might imply an original homeland from which people have scattered, and to which they might seek to return, 'routes' complicates such ideas by focusing on more mobile, multiple and transcultural geographies of home. On an individual level, 'roots figure as a referent of belonging, the position and place of a person; and routes as a referent of the lack of fixity and evolving nature of belonging' (Armbuster 2002: 30; see Box 5.2 on home and queer migrations). Rather than view home as rooted, located and bounded, and often closely tied to a remembered or imagined homeland, an emphasis on 'routes' invokes more mobile, and often deterritorialized, geographies of home that reflect transnational connections and networks. And yet, such mobility does not preclude what Avtar Brah terms a 'homing desire'. As she writes, 'the concept of diaspora offers a critique of discourses of fixed origins while taking account of a homing desire, as distinct from a desire for a "homeland". This distinction is important, not least because not all diasporas sustain an ideology of "return"' (Brah 1996: 16).

Ideas about the nation as home are often articulated through images of the homeland, as we discussed in Chapter 4, and such images are gendered and

Box 5.2   QUEER HOMES, MIGRATIONS AND 'MOTIONS OF ATTACHMENT'

In her study of queer narratives of migration, Anne-Marie Fortier describes 'motions of attachment' to unsettle – to 'queer' – the heteronormativity of home (2003; also see Fortier 2001). Many such narratives revolve around leaving the childhood, family home and seeking to find and/or create another, more accepting home. As Alan Sinfield puts it,

> Most of us are born and/or socialized into (presumably) heterosexual families. We have to move away from them, at least to some degree; and *into*, if we are lucky, the culture of a minority community. 'Home is the place you get to, not the place you came from,' it says at the end of Paul Monette's novel, *Half-way Home*. In fact, for lesbians and gay men the diasporic sense of separation and loss, so far from affording a principle of coherence for our subcultures, may actually attach to aspects of the (heterosexual) culture of our childhood, where we are no longer 'at home.' Instead of dispersing, we assemble.
>
> (2000: 103; quoted by Fortier 2003: 117; also see Brown 2000)

Drawing on Michael Brown's work, Fortier writes that '"coming out" means "moving out" of the childhood "home" and relocating oneself elsewhere, in another "home"' (Fortier 2003: 115; Brown 2000: 50). But, in contrast, Fortier analyses narratives of queer migration to question assumptions of 'home as familiarity', by decentring 'the heterosexual, familial "home" as the emblematic model of comfort, care and belonging' (115; 116). She argues instead for 'motions of attachment', whereby the home itself is mobile rather than fixed and is 're-membered' through movement and attachment:

> It is lived in motions: the motions of journeying between homes, the motions of hailing ghosts from the past, the motions of leaving or staying put, of 'moving on' or 'going back,' the

motions of cutting or adding, the motions of continual repro-
cessing of what home is/was/might have been. But 'home' is
also re-membered by attaching it, even momentarily, to a place
where we strive to *make* home and to bodies and relationships
that touch us, or have touched us, in a meaningful way.

(Fortier 2003: 131)

Fortier analyses *Night Bloom* (1998), the memoirs by US-Italian
lesbian author Mary Cappello, who writes about living in working-
class South Philadelphia, in terms of 'motions of attachment'. She
suggests that 'the diasporic home is already queer because it is
always somehow located in a space of betweenness: that it is a site of
struggle with multiple injunctions of being and "fitting in" that come
from "here" and "there"' (2003: 125). For Cappello, the home is a
site of both familiarity and estrangement: 'it is a place of disjunction,
of unbelonging, of struggles for assimilation/integration, thus a
space that *already* harbours desires for hominess' (127). Home, for
Cappello, is neither sentimentalized nor fetishized: 'The familial
home is a space that is always in construction, not only in the imagination, but in the embodied material and affective labour of women
and men' (127). Rather than regard the family home as a place that
has to be left behind, Fortier argues that Cappello's narrative articu-
lates queerness within the family home itself. Queerness in this
sense extends beyond sexuality to encompass difference and anti-
normative meanings. As Fortier explains, 'Cappello denaturalizes
any claims on the loss of home as the necessary consequence of
"coming out" and leaving home. "Becoming queer," here, is not
engendered in the movement away from home. It emerges, rather,
from the very fabric of a queer family home' (129). By unsettling the
heteronormativity of the family home, Fortier argues that 'Not only
can home be a space of multiple forms of inhabitance – queer and
others – but belonging can also be lived through attachments to
multiple "homes"' (131–2). Conceptualizing the home as queer
means recognizing 'home as a space of differences rather than
home-as-sameness' (132), which is embedded within relations of
power.

racialized, exclusionary and contested. But what happens when people leave a place that they imagine as their home and homeland? How and why are home and homeland imagined, reproduced and recast from a distance? What are the implications of a transnational homing desire for resettlement? Is it possible to return home or is 'migration a one-way trip. There is no "home" to go back to' (Hall 1987: 44, quoted in Chambers 1993: 9)? As Avtar Brah observes, some (but not all) diaspora spaces are fashioned in relation to experiences, memories and ideas about a homeland and the homing desire for some (but not all) people living in diaspora space is to return. People living in the Jewish diaspora, for example, share the 'right to return' to the state of Israel that was established in 1948. But the creation of Israel led to the dispossession and exile of Palestinians, many of whom live in refugee camps and settlements in Jordan and Lebanon, as we discuss below.

Home, homeland and migrant homecomings

For many transnational migrants, material and imaginative geographies of home are both multiple and ambiguous, revealing attachments to more than one place and the ways in which home is shaped by memories as well as everyday life in the present. In his interviews with thirty-four Barbadian Londoners, for example, John Western asked 'When you use the word "home," what are you thinking about?' (1992: 256), and interweaves his analysis with his own experiences of transnational migration and shifting ideas of home. Home, for Western, is nested within a range of scales and exists in more than one place:

> 'home' for me is, or *was* [before his father died and his widowed mother moved] a certain house on a certain Margate street; then, the Isle of Thanet; then, my county Kent; then, my country England – depending on the context. ... Having come to live half my life away from Britain, however, 'home' for me is no longer unproblematic: it is as likely to be 'Syracuse, New York,' as 'England.' 'It depends on the context' is no longer a straightforward matter of scale, of telescoping, but an admission of uncertainty and ambiguity.
>
> (256)

Western's interviewees responded to his question about home in a variety of ways:

For the British-born, and to a lesser extent for those raised in Britain, the answer is 'here, London.' Some introduce qualifications, some not. For up to half the Barbados-born of either generation, the answer is 'Barbados.' But there is much equivocation in the migrant generation. Indeed, the most common attitude among the immigrants is a striving for some *balance* between Barbados and London.

(264)

Whilst some of his interviewees sought to return to Barbados, others remained in London with their British-born children and grandchildren.

Transnational migrants often return home, either by choice or because they are forced to do so, and others migrate with an unfulfilled dream of returning home (like many people who moved to Britain from India and Pakistan in the 1950s). And yet, migrant homecomings have been, until recently, a largely neglected area of research (Harper 2005). Although estimates vary, Mark Wyman writes that 'At least one-third of the 52 million Europeans who left Europe between 1824 and 1924 returned permanently to their homelands' (2005: 16), showing the large scale of return migration. He identifies a variety of reasons for return migration: success in the new home; failure; homesickness; a call to return to take over the family farm or other property; and rejection of life overseas (21).

Transnational geographies of home are central to return migration, whereby memories of home and homeland are recast and unsettled over space and time. In her research on the British wives and children of the official elite in imperial India, for example, Georgina Gowans (2001, 2002, 2003) studies the paradoxical ways in which home and homeland were imagined and experienced over imperial space. Ideas about Britain as home and homeland lay at the heart of imperial rule in India, ensuring that the official elite would return 'home' at retirement, and that children would be sent 'home' to Britain for their health and education, far away from their parents in India. But, as Gowans shows, India also came to be imagined as home, and returning to Britain for temporary visits and on a more permanent basis at retirement was often painful and difficult. Although Britain was idealized as 'home', experiences of visiting and returning from India were much more difficult in practice. For many British children, for example, who were sent 'home' from the age of seven,

the experience was not one that was equated with notions or emotions of belonging: happiness, familiarity, stability. Perhaps unsurprisingly,

> Britain – *represented* as home (superior, enviable, familiar, stable, impe-
> rial) – was unable to live up to expectations and *experienced* as disap-
> pointing (unfamiliar, unfriendly, unstable).
>
> (2003: 432)

Many children were homesick not only for their parents but also for India.

In a different context, Alistair Thomson explores the role of homesickness in promoting return migration (2005; Hammerton and Thomson 2005). Thomson writes that around 25 per cent of the one million British people who took assisted passages to Australia in the twenty-five years after the Second World War returned home. Drawing on life-history interviews and the results from a questionnaire survey completed by more than 200 return migrants, Thomson argues that homesickness was often the main factor that led people to return to Britain: 'homesickness for people and places and ways of life in Britain' and 'a sense of "not feeling at home" in Australia' (2005a: 118) influenced the return migration of more than a third of these respondents. As Thomson shows, homesickness was 'experienced in terms of the absence or loss of different aspects of "home"' (122). Homesick migrants longed for people and places that represented home, as they missed not only family and friends, but also particular landscapes that signified home. Even though most of the postwar British migrants to Australia had lived in cities and suburbs, their homesickness often revolved around a 'homing desire' for pastoral England: 'Their Australian longing for pastoral England seems to be an over-seas manifestation of the twentieth-century cultural phenomenon by which the English middle and working classes located their national identity within rural England' (122). But, as Thomson argues, homesickness is not only about a remembered and imagined home, but also about life in the 'here and now': 'For British migrants homesickness was as much about their lives in Australia as much about their "home" in Britain' (2005a: 124). Homesickness was often felt most acutely at particular times over the life course, notably for women expecting a baby or at home on their own with young children, who missed 'the practical and emotional support of mothers and other members of the extended family back in Britain' (124). But men as well as women were homesick. Thomson identifies a more hidden form of homesickness experienced by male migrants:

> There are several examples of men who were desperate to go home but
> who blamed their return on the 'homesickness' of their wives, who in turn

explain that they felt no such thing and that their husbands were simply unable to admit to their own feelings. There are abundant clues in accounts by wives and children of the physical and emotional breakdown of their menfolk. ... These are tragic stories about the destruction of men's hopes and aspirations. These were proud working-class men whose masculine identity was primarily bound up in their professional craft and an ability to provide for the family, and who found, to their humiliation, that they were unable to sustain this identity in Australia. Barely able to admit failure, their bodies and nerves cracked up, and their only, desperate hope was to return. These men might not have labelled themselves as homesick but they had all the symptoms of the malady.

(2005a: 124–5)

Although many transnational migrants do not return home on a permanent basis and do not wish to do so, many maintain important links by sending remittances to their families and by visiting the country that they left. The global value of migrant labour remittances is estimated at $75 billion each year (Adamson 2002; also see Van Hear 2002). As well as sending money, transnational migrants are also involved in trade, consumption, and the exchange of gifts between their different homes. Ruba Salih, for example, explores the annual return visits made by many Moroccan women living in Italy, and the lengthy preparation for such visits that involves buying domestic appliances and other items such as blankets, sheets and towels, for their own use and as gifts. As she writes,

To feel 'at home' in Morocco women need to bring with them those things that constitute and represent their 'other home' in Italy. Through these commodities women display what they have become and affirm their identities contextually through objects that signal the[ir] Moroccan and Muslim belonging.

(65; for more on Muslim homes, see Campo 1991 and McCloud 1996)

Salih also shows that travelling between different homes is an ambivalent and negotiated process, particularly in terms of the relationships between husbands, wives and their extended families in Morocco.

Return visits can also be unsettling for the children of transnational migrants. In his research on 'homeland trips' for second-generation Chinese and Korean Americans, for example, Nazli Kibria (2002) explores the spatial negotiations of home and identity that take place whereby the children of transnational migrants were aware of their similarities and differences with a wider Chinese or Korean collectivity. According to Kibria,

> The sense of belonging implied by the ties of blood seemed to be overwhelmed by differences of culture as well as of nationality. Many of the second-generation Chinese and Korean Americans described gaining a heightened sense of their identity as American on the trip. If in the United States their racial identity as Asian dominated others' perception of them, in China or Korea it was their identity as American that was significant.
>
> (306–7)

In many cases, visiting their ancestral homeland enabled second-generation Chinese and Korean Americans to feel a closer affinity to the United States as home (also see Dwyer 2002, on visits to Pakistan by second-generation British Muslim schoolgirls, and further discussion on pp. 217–19 about the politics of home and identity for the children of transnational migrants).

Connections between diasporic homes and a remembered or imagined homeland are also politically important, and have often been facilitated by new media and communication technologies (see Morley 2000 for further discussion, and Box 5.3 on home and cyberspace). Fiona Adamson (2002) identifies three main ways of 'transforming home' that take place across transnational space. First, transnational migrants 'can use the political space of the transnational community as a site for the mobilization of identities, discourses and narratives that either challenge or reinforce the official hegemonic discourse of the home state regime' (156). Sometimes this political mobilization revolves around the attempt to return to a homeland or the attempt to reclaim or establish a homeland. According to Brian Keith Axel, for example, a Sikh homeland called Khalistan 'travels with the mobile imaginary of the Sikh diaspora' (Axel 2001: 199), whilst the Balkan conflicts of the 1990s saw a 'revived engagement in homeland politics' by the Croatian diaspora in the United States (Carter 2005: 56). The second way that transnational migrants work towards the 'transformation of home' is by 'networking with a variety of state and non-state actors, such as NGOs, in order to raise international awareness, thereby increasing pressures for political change in the home state' (Adamson 2002: 156), as shown by the work of Sudanese NGOs in Cairo (Häusermann Fábos 2002), Kurdish political lobbying in Germany

(Østergaard-Nielsen 2002), and political mobilization across the Kashmiri diaspora (Ellis and Khan 2002). Finally, Adamson identifies the roles of transnational migrants in mobilizing and transferring resources 'directly to actors in the home country, thus altering the local balance of resources and power' (156), as shown by organizations such as NORAID sending money from the United States to fund the Irish Republican Army. In a wide variety of ways, transnational political connections are thus closely bound up with the attempt to transform home, and reveal the material and political effects of remembering and imagining a homeland whilst living in diaspora.

Box 5.3 HOME AND CYBERSPACE

Why are personal websites called 'homepages'? What does the computer icon for 'home' represent and what form does it take? To what extent can cyberspace be interpreted as a new home-space, providing a virtual space of belonging and identification? Each of these questions shows that home is a central part of the language, imagery and politics of cyberspace. Even though many authors have written about the utopian possibilities for transcending time and space through the virtual world – and many others have critiqued the elitist exclusions of this world (for an overview, see Morley 2000) – the home remains a key site in cyberspace.

Some researchers have studied the home and cyberspace in a material sense, as shown by studies of children's use of the internet at home (Holloway and Valentine 2001) and by work on the extent to which transnational politics of home and homeland are mobilized through the internet (Staeheli *et al.* 2002). This research shows that virtual geographies stretch the home far beyond the domestic sphere, as children physically located at home can chat online to other children around the world, and as transnational and diasporic connections are forged between different homes and homelands. The internet can foster new, transnational communities of identity and belonging – a virtual geography of home – as shown by the recent diasporic interest in what it means to be an Anglo-Indian, which has been mobilized in many ways through a proliferation of websites (Blunt 2005; see, for example, www.anglo-indians.com;

and www.alphalink.com.au/~agilbert, which publishes two electronic journals about the community).

Ella Shohat writes that cyberspace is 'an embattled space for
becoming "at home" in the world' (1999: 224). Like other geographies of home, cyberspace is shaped by inclusions, exclusions and
power relations. Although the new cyber-technologies were largely
developed by and for the military, and although there are many far-
right, racist and pornographic websites, Shohat notes that such
technologies have 'at the same time become the site of activism and
grassroots organizing' for progressive politics (224). For example,
'in the Americas, as the homeland of diverse indigenous people who
became refugees in their own homeland, the new media are used to
recuperate the symbolic space of Pindorama, Land of the Condor
and Turtle Island. Such virtual spaces come to stand for an imaginary homeland' (223). As Shohat explains, the notion of an originary
and stable home is thus 'cybernetically redefined for dispossessed
nations not simply as a physical location but as a relational network
of dialogic interactions' (224).

Diasporic domestic architecture and home-making

Just as the bungalow can be interpreted as a transnational form of domestic
architecture, as discussed in Chapter 4, other house designs also reflect and
reproduce transnational geographies of home (Cairns 2004). Writing about
Carlo Levi's book *Christ stopped at Eboli*, describing life in a small southern Italian
town in the 1930s, for example, Mark Wyman observes the impact of return
migration on domestic architecture:

Levi explained how he spent his initial days wandering around the
poverty-scarred community where the only light entering most homes
came through an open door, while chickens fluttered in and out,
competing for space on the dirt floor with pigs and dogs and humans. But
here and there were other sorts of houses – painted, with a second floor
and balcony, the doorknobs bearing fancy varnish. 'Such houses
belonged to the "Americans,"' Levi wrote. In fact, the American house
became the most visible result across much of Europe of the return
migration that swept progressively in waves over the Continent. Such

houses were built by the emigrants coming back with their savings from America, or Canada, or even Argentina, where many southern Italians went.

(2005: 24–5)

The influence of migrant homecomings on domestic architecture is evident elsewhere too: 'In Portuguese towns there is the *casa francesca*, built with earnings from France, just as in another part of the world there are 'sterling' houses in parts of China, built by earnings brought home from British areas such as Hong Kong' (Wyman 2005: 27).

House styles also reflect and help to foster community identities in diaspora. So, for example, Puerto Ricans living in the South Bronx in New York built small houses or *casita* with two or three rooms and a veranda that were reminiscent of their homes in Puerto Rico (Sciorra 1996). As Sallie Westwood and Annie Phizacklea explain,

This is an importation of vernacular architecture which changes the visual and aesthetic spaces of the urban, contributing to the sense of a hybrid, diasporic spatial aesthetic. The remaking of housing, just like the dance halls and Friday night celebrations, are part of a remembering which is active, not simply a nostalgia for the familiarity of home but an attempt to make a home in a new landscape.

(2000: 63–4)

The challenges to the material forms of home wrought by migration have long been contested in many urban environments (Allon 2002). Contemporary migrations have magnified these disputes, as shown by the contested form of the 'monster house' built in Vancouver, Canada, by wealthy migrants from Hong Kong. By the 1980s and 1990s, about 8,000 people from Hong Kong arrived in Vancouver each year (Ley 1995: 195). Entering Canada as investor and entrepreneur immigrants, these people brought with them a 'new landscape aesthetic' to elite suburbs such as Shaughnessy Heights. As David Ley (1995: 191–2) describes,

This group favours new, large houses on a cleared lot, usually more than 4,000 square feet in area. The newness of the house, access to light through large windows unimpeded by vegetation, the alignment of doors, and other details of internal design, are inspired in part by the traditional metaphysic of *feng shui*. Properties on T-junctions are avoided, while traditional lucky numbers, three and eight, are much in demand. ... The home is an important opportunity to demonstrate one's appropriation of progress, one's purchase upon modernity. ... So too in their expansive homes these immigrants wish to project a successful, forward-looking, modern identity. The house, then, becomes a hybrid form, retaining certain traditional values but also proclaiming the restless modern commitment to growth and change.

This landscape aesthetic is a classic illustration of the links between imaginaries and built forms of home. It is also an example of contested ideas of home and the material forms that home should take. These new houses were built in affluent neighbourhoods in which the dominant aesthetic was as depicted in Figure 5.1 – with houses of reworked English Tudor-revival appearance, gardens of deciduous trees and shrubs, and a conservatism of interior decoration (Ley, 1995: 188). As Katharyne Mitchell puts it (and see Figure 5.2), 'The newly constructed houses contrasted vividly – in form, structure, scale, aesthetic and urban sensibility – with the historicist styles of the existing residential architecture in its picturesque suburban streetscapes' (2004: 145).

Figure 5.1 'English' landscape aesthetic in Shaughnessy, Vancouver. Reproduced courtesy of Kathleen Mee.

Figure 5.2 New housing aesthetic in 1990s Shaughnessy, Vancouver. Reproduced courtesy of Kathleen Mee.

The material form of the 'monster house' prompted bitter debates about who could claim residence – not only in the neighbourhood of Shaughnessy, but also in the Canadian nation. Public debates in planning discussions, in newspapers, and amongst builders, developers and resident associations revolved around competing definitions of home. Some existing residents emphasized home as family in contrast to perceptions about home as investment. For one resident quoted by Ley,

> We've raised our family here, sent our children to the neighbourhood schools, participated in all kinds of community events over the years. Now many of the people who own homes in the area don't live here. The homes are empty. These homes are investments, perhaps one of many. ... *We want to stress that this is a place to live not just a place to make money out of.*
>
> (quoted in Ley 1995: 197; emphasis in original)

According to Mitchell, 'the Hong Kong economic migrants quite literally brought contemporary, often paradoxical forces of dispersion, dwelling and diaspora "home" to the heretofore protected spaces of suburban Shaughnessy Heights' (2004: 143). The transnational form of the 'monster house' highlights fissures in ideals of home, as well as their connections to powerful constructions of race and ethnicity.

The material and imaginative geographies of transnational homes and homelands are manifested on a domestic scale through a wide range of home-making practices, as shown by Katie Walsh's research on British expatriates living in Dubai (Research Box 9). In many ways, transnational homes are sites of memory and can be understood as performative spaces within which both personal and inherited connections to other remembered or imagined homes are embodied, enacted and reworked (see Box 5.4 for more on home, memory and nostalgia). Such connections reveal the broader intersections of home, memory, identity and belonging across transnational space, and are materialized through, for example, domestic architecture and design, decor, furnishings and other objects within the home, and through family relationships and domestic practices.

Box 5.4   HOME, MEMORY AND NOSTALGIA

Home is often understood as a site of memory, as shown by research on material cultures and family photographs that invoke past homes in the present (Rose 2003; Tolia-Kelly 2004a, 2004b), and by research on personal and collective memories of home and homeland across transnational space (Ganguly 1992; Fortier 2000). As we discuss in this chapter, food and the social and cultural practices of cooking and eating are important in charting and maintaining a collective memory and identity, both within particular places and across wider diasporas.

Whereas sites of memory often invoke, but also extend far beyond, spaces of home, nostalgia invokes home in its very meaning. The term 'nostalgia' comes from the Greek *nostos* for return home, and *algos* for pain, and implies homesickness and a yearning for home (Chambers 1990). In Europe from the late seventeenth to the twentieth centuries, nostalgia was understood as a physical illness, but has since come to represent a state of mind (Shaw and Case 1989; Rubenstein 2001). But, by the late 1980s, 'even the pleasures of nostalgia [had] faded from memory' (Lowenthal 1989: 18) and it had become, in David Lowenthal's words, 'a topic of embarrassment and a term of abuse. Diatribe upon diatribe denounce it as reactionary, regressive, ridiculous' (20).

In many ways, a nostalgic desire for home has come to represent a wider 'desire for desire' (Stewart 1993: 23). Home becomes a temporal signifier as an imagined point of origin and return that implies a longing for an unattainable past. In her discussion of feminist fiction, for example, Roberta Rubenstein writes that

> Nostalgia encompasses something more than a yearning for literal places or actual individuals. While homesickness refers to a spatial/geographical separation, nostalgia more accurately refers to a temporal one. Even if one is able to return to the literal edifice where s/he grew up, one can never truly return to the original home of childhood, since it exists mostly as a place in the imagination.
>
> (Rubenstein 2001: 4)

Whereas sites and landscapes of memory are located and recast in the past and present, the spaces of home invoked by nostalgia remain more elusive and distant.

Alison Blunt (2005) argues that 'an antipathy towards nostalgia reflects a more pervasive and long-established "suppression of home", whereby spaces of home are located in the past rather than the present, in imaginative rather than material terms, and as points of imagined authenticity rather than as lived experience' (14). Blunt uses the term 'productive nostalgia' to focus on nostalgia as the desire for home and, rather than view this desire as apolitical or confining, to explore its liberatory potential. To do so, she studies a longing for home that was embodied and enacted in practice rather than solely in narrative or imagination, and argues that a nostalgic desire for home is oriented towards the present and the future as well as the past.

Transnational connections are also materialized within the home through a range of other home-making practices. Transnational home-making is a gendered practice, and represents part of a broader process of 'feminising the diaspora' (Walter 2001: 11) that is characterized by the migration of women and the domestic symbols often used to represent resettlement. In her research on women living in the Irish diaspora in Britain and the United States, Bronwen Walter traces identities of placement and displacement, which are mapped onto different geographies of home. Identities of *displacement* invoke Ireland as home, and revolve around memories of growing up in Ireland, return visits, a sense of national belonging, and shared memories and stories that connect people across diasporic space. In contrast, identities of *placement* are bound up with homes and families outside Ireland and Irish community life in diaspora. In both cases, Walter explores the gendered practices of diasporic home-making, and writes that

> Diasporic women are ... placed in a paradoxical relationship to home. It can be a source of containment and fixity, rendering women invisible, and linking them to the mundane and routine. But it can also be the basis for challenging dominant cultures both outside and inside their own ethnic group. Issues of identity and belonging are thus raised for diasporic women by their relationships with home.
>
> (2001: 197)

In other words, diasporic homes are sites of both containment and potential liberation for women. Two further examples illustrate this point. First, Susan Thompson describes the 'power of home' for migrant and first-generation Australian women from the Arabic, Greek and Vietnamese communities (1994), whereby the home is a place 'where difference can be displayed and acted out' (37) by speaking languages other than English, and by fostering cultural identities through domestic interiors and material cultures. Second, drawing on interviews with Anglo-Indian women who migrated to Britain in the late 1940s and 1950s, Alison Blunt (2005) explores the familiarity and unfamiliarity of home. Unlike many other people who migrated to Britain from the Indian sub-continent, Anglo-Indians thought that they would be settling in a more familiar culture and lifestyle than they anticipated in independent India. But their domestic life on resettlement proved to be unfamiliar in many ways, particularly because of the small size of houses and because Anglo-Indians had to do domestic work without servants. The domestic challenges of resettlement were felt most acutely by women, particularly in terms of learning how to shop, cook and clean for the first time. Remembering how they met such domestic challenges provides an important narrative of survival and success for Anglo-Indian women in Britain today.

Rather than view diasporic home-making as the reproduction of static notions of tradition and culture, it is important to think about the ways in which it is a dynamic and transformative process shaped by the mixing and reworking of traditions and cultures. In her research on South Asian women from South Asia and East Africa living in London, for example, Divya Tolia-Kelly studies the importance of visual and material cultures in shaping diasporic homes. Tolia-Kelly argues that the visual and material cultures that shape 'new textures of home' in London 'are shot through with memory of "other" spaces of being' (676). Studying photographs, other images and mementoes that resonate with memories of home prior to migration, Tolia-Kelly shows that 'visual and material cultures are prismatic devices which import "other" landscapes into the British one, and thereby shift notions of Britishness, and British domestic landscapes' (678). For example, a photograph of Bismarck Rocks (later named Mwanza Rocks) on Lake Victoria is a powerfully evocative image of home for Sheetal, who grew up in Tanzania in the 1960s and 1970s (Figure 5.3). Displayed in her London bedroom, this photograph transports Sheetal back to a particular place and time, whereby 'the domestic landscape in Britain is contextualised through Tanzania' (681). In contrast, a particularly resonant image displayed by Shilpa in her London home is called 'Boy fishing' (Figure 5.4). Tolia-Kelly describes this picture as a 'sugar sweet greeting-card image of a rural idyll, a childhood idyll in fact, certainly aesthetically situated in a European or English landscape' (684). But, for Shilpa, this image evokes her childhood memories of her uncle's home by a river in Uganda:

this image is made meaningful in her flat in Harlesden. It is positioned in Britain as a testament to the landscape of dreamy childhood days in Uganda of a luxurious home, a luscious countryscape and the pleasure of

Figure 5.3 Bismarck Rocks. Reproduced courtesy of Divya Tolia-Kelly.

Figure 5.4   Boy fishing. Reproduced courtesy of Divya Tolia-Kelly.

free roaming. ... The image of an English pastoral in this story is cross-cultural; it is translated as an aspiration toward a picturesque scene, which embodies pleasure, peace and a sense of innocent childhood pastimes.

(684)

In addition to visual and material cultures, the preparation and consumption of food are also important practices of diasporic home-making and clearly reflect the mixing and reworking of traditions and cultures (Hage 1997; Kneafesy and Cox 2002). For example, many Anglo-Indians who lived in India before Independence imagined that they were part of a British imperial diaspora. But whilst their home-life was in many ways more western than Indian – as shown by their home furnishings and embodied domestic practices, such as using cutlery rather than eating by hand – the food that they ate reflected both western and Indian influences. Anglo-Indians usually ate curry and rice at lunch but a western-style breakfast, tea and 'side dish' in the evening. As an Anglo-Indian who lives in India explains, 'we picked up the traits of ... cakes for tea and crumpets [and] we also picked up the spicy curries for lunch' (quoted in Blunt 2005: 54). A 'masala steak' represents the hybridity of Anglo-Indian cuisine. In the words of an Anglo-Indian who now lives in Australia, 'A steak is British and masala [is Indian], to give it a bit of difference. So that became the culture of the community too ... made up dishes that suited both tastes' (quoted in Blunt 2005: 54). In her research on Moroccan women living in Italy, Ruba Salih (2002) also explores the home as a distinctive site reflecting 'double belonging' and a 'plural identity' (56), both in terms of the visual and material cultures of home and through the preparation and consumption of food. As she writes,

> Typical cheap, popular Italian furniture is displayed together with objects recalling the Moroccan and Muslim world such as covers for sofas, pictures showing Quranic writings on the walls and, in some cases, calendars arriving from France with dates of Islamic celebrations and feasts or posters of Moroccan women wearing the traditional dresses of different areas of Morocco. ... Together with Italian food, spices brought back from Morocco and particular ingredients are kept in big quantities. Some of them are used during Ramadan or on other special occasions. The typical Moroccan terracotta cooking pots used to prepare *tajin* were also present in many of the homes I visited in Italy.
>
> (56)

Food's relationship to transnational forms of home is multiple. George Morgan (2005) and colleagues argue that for migrants in outer Sydney, 'their gardens are often places of connections not only to homeland but also to Australia and other cultures' (95). Food that reminded them of their homeland was produced at home. As one woman recalled,

> When one goes out into the backyard and sees the Cuban sugar cane, it reminds you of your country, of your life in Cuba, where you were born. This is the sugar cane that I had on my father's farm ... The guavas are Cuban guavas, because here there are no guavas. And the bananas are like we had next to our house there, and we picked them by the hand and we used to fry them at home. And you feel as if this was your country ...
>
> (Morgan *et al.* 2005: 96)

Hybrid homes

The hybridity of diasporic home-making is particularly evident across different generations. For example, the British Muslim schoolgirls interviewed by Claire Dwyer, whose parents had been born in Pakistan, articulate their British Asian or British Pakistani identities in relation to transnational geographies of home (Dwyer 2002). Rather than view home and identity as static, fixed and singular, these articulations reveal their dynamic multiplicity in relation to different places. As Dwyer explains, 'In the process of making home in Britain, other places are also made home as identities are made, re-made and negotiated. … [T]he evocation of Pakistan as "home" suggests not a return to a mythical or lost "homeland" or roots but instead a symbolic process of making home' (197). Building on work by Stuart Hall (1996), Dwyer unsettles the distinctions between 'home' and 'abroad', 'here' and 'there', as the young women 'negotiate belongings to several different "homes" at one and the same time' (198). But such multiple belongings can be painful and difficult. In her research on second-generation Filipino students in the United States, for example, Diane Wolf (2002) explores what she terms 'emotional transnationalism'. Like the schoolgirls in Dwyer's study, the Filipino students interviewed by Wolf described their attachment to their parents' homeland: 'despite place of birth, and whether or not they had lived in or visited the Philippines, the students, in their references to "home", always meant the greater Home across the ocean' (263). Their parents' memories of the Philippines as 'Home' shapes everyday life for second-generation Filipinos in the United States because

> This Home is morally superior to the home they now inhabit and constitutes the foundation for judging behaviors as proper, appropriate, or shameful. … As [the children of immigrants] manage and inhabit multiple cultural and ideological zones, the resulting emotional transnationalism constantly juxtaposes what they do at home against what is done at Home. While this may offer the security of a source of identity, it also creates tensions, confusion, and contradictory messages that … can lead to intense alienation and despair among some. The son who is told, 'Just because you live here in America, don't be influenced by everything you see, because you are Filipino and should know who you are and where you come from,' is being told to differentiate between home and Home.
>
> (285)

Other studies of 'growing up global' as 'Third Culture Kids' (Eidse and Sichel 2004; Pollock and Van Reken 1999) reveal the mobilization of home across generations and across transnational space. In an edited volume of life writings about unrooted childhoods, for example, Faith Eidse and Nina Sichel explain that the children of 'educators, international businesspeople, foreign service attachés, missionaries, military personnel … shuttle back and forth between nations, languages, cultures, and loyalties' (1). Such 'nomadic children' not only 'often feel as though they are citizens of the world and must grow to define home for themselves' (1) but also 'often question the whole concept of home, never feeling that they quite belong anywhere. They wonder who they are and whether they can settle anywhere permanently' (4). In an autobiographical essay, for example, Nina Sichel describes being born in the United States to an American mother and a German father, being brought up in Venezuela, and returning to New York every summer. As she writes, home is not a 'real place', but rather a 'shifting definition' (185). Home is similarly transnational and shifting for Adrian Carton (2004), the son of an Anglo-Indian father and a British mother, who grew up in Britain before migrating to Australia. In an essay that describes his childhood memories of Britain, resettlement in Australia, and his first visit to India, Carton describes himself as a transnational subject with multiple senses of home who is a 'passionate occupant of the transnational transit lounge' (for a classic memoir of home and transnational migration, see Hoffman 1989, and for fictional accounts of transnational homes and identities, see, for example, The Joy Luck Club by Amy Tan, Brick Lane by Monica Ali, The Buddha of Suburbia by Hanif Kureishi, and 26a by Diana Evans. Also see Research Box 10 by Rachel Hughes on home and migration in Jamaica Kincaid's novels).

HOME, EXILE AND ASYLUM

In this section we turn to focus on meanings and experiences of home, various home-making practices, and the connections between home and homeland, for exiles, refugees and asylum seekers who have been forced to migrate because of war, persecution or dispossession. The term 'exile' 'suggests a painful or punitive banishment from one's homeland. Though it can be voluntary or involuntary, internal or external, exile generally implies a fact of trauma, an imminent danger, usually political, that makes the home no longer safely habitable' (Peters 1999: 19; also see Øverland 2005). The definition of a 'refugee' in international law dates from the 1951 United Nations Convention relating to the Status of Refugees, which states that a refugee is any person who:

> owing to a well-founded fear of being persecuted for reasons of race, religion, nationality, membership of a particular social group or political opinion, is outside the country of his [sic] nationality and is unable to or, owing to such a fear, is unwilling to avail himself of the protection of the

country; or who, not having a nationality and being outside the country of his former habitual residence as a result of such events, is unable to or, owing to such a fear, is unwilling to return to it.

(Quoted in Bloch 2002: 7)

An asylum seeker 'is a person who is seeking asylum on the basis of his or her claim to be a refugee' (Bloch 2002: 8). As Table 5.1 shows, the Office of the United Nations High Commissioner for Refugees (UNHCR) estimates that there were more than 17 million asylum seekers, refugees and others of concern in 2004.

Many exiles, refugees and asylum seekers have witnessed the destruction of their homes through their experience of 'extreme domicide' (Porteous and Smith 2001; see Box 4.6). The Israeli Army, for example, destroyed 385 of the 475 Palestinian villages that existed before 1948 (Dorai 2002: 95). Palestinians living in refugee camps and settlements in Jordan and Lebanon, as well as the wider diaspora living in Europe and North America, have lost their homes and their homeland. Many people from Croatia and Bosnia-Herzegovina became refugees and asylum seekers as a result of war in the former Yugoslavia in the 1990s. As Maja Povrzanovic Frykman writes, 'It is not only their concepts of homeland that have been transformed, but also their homes in the most basic, physical sense. From sites of personal control, they were transformed into sites of danger and destruction. ... People were forced to leave their homes in response to threat, fears, military orders and violent attacks. Many homes literally ceased to exist' (2002: 118; also see Carter 2005).

Table 5.1 Estimated number of asylum seekers, refugees and others of concern to UNHCR, 1 January 2004

Asia	6,187,800
Africa	4,285,100
Europe	4,242,300
Latin America and Caribbean	1,316,400
Northern America	978,100
Oceania	74,400
TOTAL	17,084,100

Source: www.unhcr.ch/cgi-bin/texis/vtx/basics

Home, housing and shelter

People who have been forced to flee their homes and homelands live in a wide range of locations and circumstances. Whilst many remain homeless, others live in refugee camps, particularly in Africa, or in public housing, hostels, reception centres or detention centres in Europe and Australia. Housing and settlement policies for asylum seekers and refugees vary from place to place. In Kenya, for example, more than 250,000 refugees lived in camps at the end of 1994, which were generally located 'at the geographical and economic margins of the country' (Hyndman 2000: 87). By living in such camps, refugees exchanged 'the rights and entitlements of citizenship for safety' (87). As Jennifer Hyndman explains,

> The Kenyan government insists that all prima facie refugees – whose status is designated by UNHCR, not by refugee law or by the Kenyan government – live in camps where they are prohibited from seeking employment or moving around the country. Instead, they are provided with food twice a month, basic medical services, primary schooling, and some housing materials, most of which are paid for by donor governments in Japan, Europe, and North America.
>
> (87–8; also see Bookman 2002)

Like refugee camps, detention centres throughout the world are profoundly unhomely places. Drawing on Augé's work on 'non-places' and Agamben's ideas about 'spaces of exception', Bülent Diken writes that

> Indefinite imprisonment, not being told of one's rights, delays in responses to requests for legal assistance, being held in isolation from other parts of the detention center, the use of force, and poor general conditions regarding food, medical services, privacy, sleeping arrangements, the level of personal security, and education and recreation facilities: these are the most common characteristics of life in most detention centers all over the world.
>
> (2004: 94)

Reception centres for asylum seekers are also unhomely places. In her research on a reception centre in Rotterdam, the Netherlands, Hilje van der Horst (2004) draws on interviews with residents and argues that 'Of the meanings attributed to home, autonomy and the possibility to lead your life in accordance with certain cultural customs are most missed in the reception centre'

(43). Unlike most reception centres in the Netherlands, which are rural and remote, the centre studied by van der Horst is located in 'the middle of a lively, ethnically mixed neighbourhood in Rotterdam' (42). The converted hostel houses about 355 asylum seekers, and is also used by various professional organizations, rendering it 'a public more than a private place' (42). Almost all of the interviewees had lived there for more than a year, and sometimes for longer than three years. As van der Horst explains, the lack of autonomy was a key part of the unhomeliness of the centre:

> Visitors have to leave before a certain time in the evening, loud music is not allowed and cleaning is checked regularly. In addition, the childlike treatment, such as giving pocket money instead of salary or welfare, supervision of visitors, checks on tidiness, takes away a few taken for granted privileges of adult life and the ability to lead your own life.
>
> (43)

Furthermore, the difficulties of upholding certain cultural customs make the centre an unhomely place, both in terms of the absence of separate spaces for men and women, and the difficulties of maintaining traditions of hospitality and 'suitable relations between family members' (43). For many asylum seekers, the relationships between home and family were disrupted by the lack of space in the reception centre:

> A considerable number of asylum seekers come with family members. When possible they are placed together in rooms. It is a major complaint of many of the inhabitants when strangers sleep in the same space, but also when family members share a small room. Good manners between family members and rules of intergenerational respect are disturbed in the centre because family members are forced to live together in a small space.
>
> (43–4)

In contrast, another important aspect of home as an autonomous and culturally important space – the ability to choose what to eat, and to prepare the food – has significantly improved over recent years: 'It was an important increase in autonomy for people living in the reception centre when in 1997 kitchens were installed and money was given to the residents to buy food. The possibility to make one's own food was very important for the well being of many residents' (43).

Government policies towards immigration, asylum and housing for asylum seekers and refugees vary from place to place. William Walters' investigation of notions of security and citizenship in the British government's 2002 White Paper on immigration entitled 'Secure Borders, Safe Haven' demonstrates the links between transnational migration, especially of asylum seekers, and nation as home (Walters 2004; also see Chapter 4 on domopolitics). Walters argues that

> Insecurity is bound up with themes of mobility: it is the movement, the circulation, the presence of unauthorized bodies which have violated the borders of the nation-state. But insecurity is connected at the same time to criminality, with activities occupying a domain outside the law, transgressing 'our' values, 'our' way of life. We are confronted with illegal acts: it's almost as though our response to them needs no further explanation. Domopolitics: 'our' homes are at risk.
>
> (247)

Like the politics of homeland security, the politics of social security – in this case, immigration policy – depend upon, and perpetuate, a normative assumption of the nation-as-home. Both 'foreign' and 'mobile' people are seen to threaten the borders of the nation-as-home, as shown by the very language used – '*illegal* immigration, people *trafficking*, *abuse* of the asylum system' (247) – which serves to criminalize asylum seekers. For Walters, the White Paper is an example of domopolitics, which 'aspires to govern the state like a home' (237). Domopolitics 'mobilizes images of home, a natural order of states and people, of us and them' in an attempt 'to contain citizenship, to uphold a certain statist conception of citizenship in the face of social forces that are tracing out other cultural and political possibilities' (256). It does so, in part, by casting 'the mobilities of survival and the assertion of a right to settle as "illegal" and "dangerous,"' (256) and, as a result, seeks to strengthen state borders to protect the homeland from such mobile insecurities (see Box 5.5 for alternative views).

Box 5.5   HOMELY CONNECTIONS IN A GLOBAL WORLD

Ideological and political linkages between home, nation and terror are becoming more commonplace, encapsulated in Walters' notion of 'domopolitics' in which a bounded conception of home is articulated to create an 'us' and 'them' of the nation. But homely connections – of family, familiarity and belonging – that are porous and multi-scalar are also being used to rethink affinity in a global world that moves beyond us and them. In an evocative piece, Nicola Evans uses the popular 'six degrees of separation' notion to turn 'networks of conspiracy into connections of conviviality and enlarging the spaces of the world into which we are moved to enter and in which we might feel safe' (146). 'Six degree chains' are based on the likelihood that two strangers from different parts of the world might have a friend or acquaintance in common. A study in the 1960s, and subsequently replicated mathematically, found that 'two strangers living in different parts of a continent inhabited by more than 200 million people were connected by only 5.5 intermediaries' (Evans 2005: 140). These chains of acquaintance were popularized in the film *Six Degrees of Separation*, and a 'parlour' game is played in the United States and UK in which people try to connect themselves to others through six acquaintances or less. Like Walters, Evans recognizes that recent campaigns to promote public awareness of terrorism have highlighted the presence of terror at all scales, including home, and a reinscription of the stranger as someone outside, who does not belong, in any of these homes. Such campaigns, and other trends, render it more difficult to feel 'at home in the world': to feel safe in a world of strangers (144; also see Ahmed 2000). Six-degree chains are important for Evans because they 'make small worlds large by bringing more people into our mental imaginary, enriching our perceptions of the kinds of people with whom we might be sharing a connection' (142). Although Evans doesn't use these terms, she is explaining a process whereby the world is imagined to be more homely, constructed through processes of connection and affinity rather than otherness. In her terms, such small-world encounters are moments

> When our sense of the size of things, the largeness of the world, the smallness of our place in it, the sense of what is close to home or distant is completely confounded, short-circuited by a connection that appears magically to have occurred against all reasonable odds.
>
> (140)

In the UK, the housing rights of asylum seekers have become more restricted since the 1980s. In 2000, the government introduced a policy of dispersal to ease pressure on housing and other services in London and the South East. According to the Housing Associations' Charitable Trust (hact),

> Applicants for asylum who need accommodation are now sent without choice to areas where housing is available. Most dispersal or 'cluster' areas are poor neighbourhoods where services, such as health, are often inadequate. Many areas have no recent track record of migrant groups settling. All of these factors, as well as poor preparation of dispersal areas, has resulted in tensions in many neighbourhoods, as well as difficulties for refugees in gaining access to housing, benefits etc. after getting leave to remain or refugee status.
>
> (www.hact.org.uk; also see the websites of Shelter and the Refugee Housing Association for further information about policy in the UK: http://england.shelter.org.uk and www.refugeehousing.org.uk)

Through its Refugee Housing Integration Programme, hact 'aims to achieve more integrated neighbourhoods in which there is an increase in the amount and quality of housing available to refugees' (www.hact.org.uk).

But, as we have argued throughout the book, the provision of housing does not necessarily mean that people feel at home. In her research on the settlement of refugees from Somalia, Sri Lanka and the Democratic Republic of Congo living in the east London Borough of Newham, for example, Alice Bloch (2002) writes that 'Most refugees do not choose to come to Britain, do not settle here and do not want to stay' (160). Bloch found that 43 per cent of her respondents saw Britain as home. This varied by country of origin, the basis of the asylum claim, and the security of immigration status. According to Bloch's survey, 7 in 10 Tamils from Sri Lanka regarded Britain as home, compared to 4 in 10 Somalis and 2 in 10 Congolese. Identifying Britain as home was also strongest amongst those seeking asylum on the basis of race, and amongst those who had the most secure immigration status. As Table 5.2 shows, the refugees interviewed by Bloch gave a wide range of reasons for feeling either at home or not at home in Britain.

Home-making

Like many other transnational migrants, the everyday lives of many exiles, asylum seekers and refugees are closely shaped by memories of home and

Table 5.2   Refugees' reasons why Britain is seen as home or not as home

	Frequencies
Do not see Britain as home	
Different culture	28
Have my own country / home is where I was born	19
Racism	12
Cannot as a refugee	12
No decision on asylum case / no refugee status	11
Climate	10
Only here temporarily / will go home	9
Britain second home	6
Different language	6
Limited rights	5
Lonely / miss family	4
Different food	3
Other	2
See Britain as home	
Living here now	18
Feel settled	14
Free / safe country	12
Second home / future home	10
No choice	5
No plans to move	3
Taken care of here	3
British government has a responsibility	3
Know nothing else	1
Other	2
Base number = 178	

Source:Bloch 2002: 143–4.

homeland. In Palestinian refugee camps in Jordan and Lebanon, for example, memories of the home village inform everyday life and a sense of identity across different generations. As Mohammed Kamel Dorai writes,

> It is always astonishing to hear children of the third exiled generation name the village of their grandparents when asked where they come from. Their answers are all the more astonishing given that most of the villages were completely destroyed long ago, and exist nowhere but in the memories of the refugees. In the camps, photographs, and even small gardens, evoke memories of villages in Palestine.
>
> (2002: 93)

Dorai argues that the refugee camps perpetuate the Palestinian homeland in exile, both in terms of materializing memories of the exodus after 1948 and through settlement patterns. As he writes,

> Family memories are transmitted orally through narrations of the exodus and recalling a past life. Life in the camps is thus justified by the history of the exodus and the camps therefore assume a new meaning for their inhabitants. The camps have become more simply than somewhere to survive, they also symbolize the exodus. It is as if Palestinians brought with them a piece of their land and deposited it in the camps, thus recreating a part of Palestine. ... In a more literal sense, the camps often spatially recall settlement patterns in Palestine, as refugees have tended to settle according to their village or area of origin. Different parts of the refugee camps often bear the names of the villages of origin of the refugees in Palestine.
>
> (Dorai 2002: 93–4)

Dorai explains that everyday life in refugee camps fosters 'the permanence of the homeland' (94) and that the camps are places not only to recall the homeland but also to wait to return home.

Returning home

Returning home is a difficult and dangerous process for many exiles, asylum seekers and refugees. In her research with refugees in Newham, Alice Bloch found that 71 per cent of respondents would like to return home given the right circumstances, compared to 19 per cent who said that they might want to return home and 10 per cent who definitely did not want to return to their country of origin (2002: 145). Peace and/or democracy were the determining factors for almost all respondents who wanted to return home. The reasons given by the 10 per cent who did not want to return included 'the length of time already spent in Britain and having children who do not know the homeland and might therefore experience difficulties on return' (145).

Repatriation does not necessarily imply a return home. In an ethnographic study of refugee repatriation to Ethiopia, Laura Hammond challenges the widely held belief 'that refugees who return to their country of origin are necessarily people "going home"' (2004: 3). An estimated 200,000 people fled from the region of Tigray to Sudan in 1984 because of civil war. Hammond focuses on the repatriation of Tigrayan refugees to a settlement called Ada Bai from 1993, and traces the ways in which returnees 'set about transforming an unknown and anonymous space first into a personalized place and finally into a "home"' (3). Although Ada Bai and its surroundings had been assigned to Tigray Region, and the repatriated refugees were thus described as returning to their homeland, it was a new and unfamiliar place. Most notably, Ada Bai was located in the lowlands rather than the highlands, necessitating social and economic changes in the lives of the returning refugees.

Rather than view home as a physical place to return to, Hammond understands home 'as the conceptual and affective space in which community, identity, and political and cultural membership intersect. In this sense, home is a variable term, one that can be transformed, newly invented, and developed in relation to circumstances in which people find themselves or choose to place themselves' (10; also see Pilkington and Flynn 1999 on the movement of Russians from other republics of the former Soviet Union, and Eastmond and Öjendal 1999 on the return to Cambodia). As we have argued throughout the book, home is not a fixed and static location, but is rather produced and recast through a range of home-making practices that bind the material and imaginative geographies of home closely together and exist over a range of scales. For the Tigrayan refugees who settled at Ada Bai, such home-making practices involved strategies of emplacement to make the returnee settlement feel like home. Such strategies involved 'the interworking of place, identity, and practice in such a way as to generate a relationship of belonging between person and place. Emplacement involved the gradual expansion of places that people considered to be familiar and safe from the raw material of a space that was unfamiliar and dangerous' (Hammond 2004: 83). Home-making through emplacement occurred both within and beyond the scale of the house. As Hammond explains,

> Whereas the process of home-making in Ada Bai involved re-creating a house that was historically, physically, aesthetically, and practically similar to the dwelling in the refugee camp, place-making in the wider physical environment was more a process of fashioning a new space to fit the practical and spiritual requirements of everyday life.
>
> (99)

For Tigrayans, home relates to the idea of *hager seb*, which means 'the place of one's people' (106), and may take generations to establish. Although aid agencies and governments may view repatriation and homecoming as synonymous for returning refugees, and expect the process to take only one or two years and be largely unproblematic, Hammond shows that repatriation and homecoming are not the same, and that the ability to feel at home on return is a long and challenging process.

The difficulties of returning home are also powerfully articulated on a personal level in autobiographical writings. In Edmunds Bunkše's book on 'the geography within my heart and soul', for example, 'home and road' and the difficulties of returning home are important themes (2004: 5; also see Box 1.3 on the poetics of home). Born in Latvia, Bunkše and his family were forced to flee west when Russian forces invaded in 1944. Aged nine, Bunkše lived with his family in a displaced persons camp for five years before migrating to the United States in 1950: 'Since I was dispossessed of my homeland at an early age, my life has been dominated by a conscious awareness of home and homelessness, of being on the road both literally and metaphorically' (6). Bunkše describes his 'new homeland' in the United States as 'both home and road, for the Baltic Sea and the hills and meadows of Latvia, together with its northern solstices and my native language and mentality, continue to tug at me' (7). In 1990, Bunkše returned to Latvia for the first time, and lived there for six months on a Fulbright scholarship: 'almost forty-four years after our flight from Latvia a journey home had begun. The road home was open for the first time' (46). He describes in moving detail the fear and pain he experienced on his return:

> I found that it is impossible for me to go home. That is not an uncommon experience in this postmodern world, but it took a journey to my homeland to learn that my ideas about that land and its people were only

illusions. ... I have found that a home is more than a national anthem, a beautiful city center, patches of beautiful rural and "natural" landscapes, a stormy, sometimes utterly calm sea and white, amber-bearing beaches, some festivals, a few friends and relatives. Nostalgia for a homeland cannot overcome the passage of time. The journey home was wrenching.

(56)

Bunkše's most painful experience was visiting his grandmother's farm, which had by then been subdivided to house four Russian families. He describes in powerful prose the 'life-altering shock' he experienced when a couple living there was told that he belonged to the family of the former owners:

The woman immediately began screaming and running in a circle, wringing her hands. She thought I had come to reclaim the farm and to evict her, and, still screaming, she tried to show me the improvements they had made. The screams and the sight of the water pail [to catch water leaking through the roof] were too much for me. I fled outside, ran to the other side of the house, and doubled over to vomit. But no vomit came, only dry heaves.

(39–40)

For Bunkše, and for many other people living in exile, it is impossible to return home.

TRANSNATIONAL GEOGRAPHIES OF DOMESTIC WORK

One aspect of contemporary transnational migration has multi-layered relevance to a critical geography of home: the movement of international domestic workers. The number of international domestic workers is almost impossible to even estimate due to a lack of available and accurate data and because much of this work is unregistered and unregulated (Ramirez-Machado 2004: 1). It is known that those who work as nannies, au pairs, cleaners and maids in countries other than those in which they were born are almost exclusively women. Geographically the movement is primarily from the poorer 'south' to the wealthier 'north', with countries such as the Philippines being a major 'supplier' of global domestic workers. The re-emergence of widespread paid domestic work, though on a global rather than national scale, is the result of a number of factors. Though these factors are complex, we will rehearse them here as a way of providing a context for the discussion that follows. First is a general increase in the transnational movement of people for work purposes, and women make up more than half of contemporary global labour migration (Castles and Miller 2003: 9). More specifically, though, transnational domestic work is connected to persistent disparities in world incomes and limited opportunities for high incomes in the developing world. In the developed world, rising incomes and the increasing full-time employment of women, especially in professional fields, combined with perceived shortages and high cost of domestic workers locally, have led to searches for domestic workers from abroad. In the developing world, the income offered by working abroad for short or long periods has become part of family strategies to survive and provide improved opportunities for future generations. For example, in the 1990s a middle-class Filipino woman could earn around US$176 a month in the Philippines. If she instead was employed as a domestic worker overseas, her monthly income would rise to US$200 in Singapore, US$700 in Italy or US$1,400 in Los Angeles (Hochschild 2003: 18). Both these factors are overlain by various state policies and programmes that encourage transnational domestic work. Recipient countries such as Canada and the United Kingdom have instituted special visa categories and entry requirements to simplify and facilitate the migration of domestic workers. In sending countries such as the Philippines, specific programmes and agencies of government are charged with organizing and managing the more than seven million Filipinos who work abroad, alongside numerous migration agents (Castles and Miller 2003: 168). Global domestic work has attracted a fair amount of critical attention from geographers and feminists, in terms of its implications for global inequalities (Hochschild 2003), constructions of gender and race (Pratt 1997) and conceptualizations of globalization (Nagar et al. 2002) (see Box 5.6 on the political organizing of domestic workers). In terms of a critical geography of home, the globalization of domestic work amplifies the porosity and stretching of home in the contemporary world, as we demonstrate in this section.

Box 5.6 LEGAL RIGHTS AND POLITICAL ORGANIZING OF DOMESTIC WORKERS

Recent national and international campaigns have highlighted the poor employment conditions experienced by domestic workers worldwide. These include long hours of work, lack of rest periods and days off, as well as the lack of privacy we document in this chapter. Surveys of domestic workers have also found widespread instances of emotional and physical abuse. A British survey conducted in the late 1990s found that 87 per cent of overseas domestic workers had been subject to psychological abuse; 38 per cent did not get food regularly; and 58 per cent were paid less than agreed to in their contract (KALAYAAN/CFMW Large Sample Statistics presented to UK Select Committee on Home Affairs, 12 May 1998). Because domestic work occurs in a home, formal regulation of the conditions of employment of domestic workers has proved difficult and problematic. In particular, presumptions of the private nature of both the workplace and the nature of the employment have underpinned a reluctance to regulate as well as the nature of regulation. According to a recent International Labor Office summary:

> Of the national laws analysed, quite a considerable number of countries (nine) exclude domestic workers from general labour laws applicable to other categories of workers. In several countries (19), the law does not make express reference to them. Some others (20) have set up specific regulations on domestic workers in their Labour Code, or equivalent legislation (i.e. general law on working conditions, law on the working environment, Labour Protection Act, etc.). However, in some cases, this may simply mean that the general labour legislation does not apply in its entirety to domestic workers. Finally, in another group of countries (19), the specificity of domestic work has been recognized by the enactment of special laws dealing with this category of worker. Regardless of the manner in which domestic work is regulated by national laws, it may be said that, in general terms, standards on domestic work fall below labour standards set for other categories of workers.
>
> (Ramirez-Machado 2004: 64)

Partly because of the relative lack of formal regulation, various organizations and campaigns working for the rights of domestic workers have emerged, some of which we list in Table 5.3. These organizations not only act as a point of contact for domestic workers concerned about their rights, they also work to publicize the situations of domestic workers since they remain invisible. There are two characteristics of these campaigns especially relevant to the critical geography of home we have developed in this book. First, they extensively use the internet in both their advocacy and support functions, emblematic of the potential of cyberspace to forge diasporic connections (see Box 5.3). Second, these campaigns are multiscalar, recognizing *both* the geographical dispersal of domestic workers and their embeddedness in local contexts. For example, the 'Break the Chain Campaign' is based in Washington, DC, and specifically represents those working on special visas issued for household employees of diplomats and employees of international agencies. In contrast, the Commission for Filipino Migrant Workers works with and through the Filipino diaspora scattered across Europe.

Table 5.3   Transnational domestic worker organizations and campaigns

Commission for Filipino Migrant Workers: 'A House is not a Home' programme (www.cfmw.org/files/migrantwomen.htm)

Human Rights Watch: 'Hidden in the Home' (www.hrw.org/reports/2001/usadom/usadom0501-01.htm)

KALAYAAN – Justice for Overseas Domestic Workers (http://ourworld.compuserve.com/homepages/kalayaan/home.htm)

Labour and Migration Task Force, Asia Pacific Forum on Women, Law and Development (www.apwld.org/lm.htm): focus on Domestic Work 2003–5. Legal recognition and protection for migrant domestic workers

Break the Chain Campaign; The Institute for Policy Studies, USA (www.ips-dc.org/campaign)

United Domestic Workers of America: www.udwa.org

Domestic Workers United: www.caaav.org

Transnational migration influences lived experiences of home in the west in a multitude of ways. In Chapter 3 we documented how the performance of unpaid domestic work, primarily by women, was a key home-making practice. Transnational migration magnifies the home as a site of paid rather than unpaid work, a simultaneously public and private space. Arlie Hochschild (2003) further argues that contemporary paid domestic work fundamentally changes the nature of domestic labour through its commodification and geographical shifting of the work of caring. According to Hochschild, paid domestic work is essentially about outsourcing the emotional and social work of the home – raising children, establishing emotional and social ties within and beyond the home. In the homes of affluent westerners, the caring aspects of home-making are now performed by women from the developing world. The employment of global domestic workers also renders the home a site for the circulation and negotiation of racialized and cross-cultural identities. Often employers operate with stereotypical notions of Filipino nannies, for example, considering them as 'natural' carers (see Pratt 1997).

Lived experiences of home for domestic workers stress home as a workplace and as a site in which their autonomy and control is limited. Nicole Constable's research on Filipino domestic workers in Hong Kong revealed rules and duties typically associated with workplaces (Constable 2003). Workers' use of time and activities were carefully monitored, sometimes through detailed daily schedules. For one worker, this included instructions on what to clean at what time, how to prepare dinner, when to feed the dog and when to rest (for one hour between 1 and 2 pm) (Constable 2003: 124). Moreover, as we highlighted in Chapter 3, the notion of home as a place of personal freedom was challenged daily. Home for many of these women was a place of curfews, surveillance in the form of constant phone calls from employers or sometimes the presence of a family member in the house, the compulsory wearing of a uniform and other dictates about appearance. Home was not a sphere for oneself, independent judgement, or personal time:

> Regardless of whether the floors or windows appear clean, they must be cleaned at the scheduled time. In most cases, workers are not allowed to judge for themselves whether a job needs to be done; they are merely required to follow the schedule. Many domestic workers are not permitted to write letters or take care of their own business on workdays or during work hours regardless of whether they have completed their duties.
>
> (Constable 2003: 122)

Home hence became a site of tension and conflict, especially around food. Whereas food preparation and consumption is recognized in other contexts as playing an important role in constituting senses of belonging and connections among household members and over diaspora space (DeVault 1991; Bell and Valentine 1997), for domestic workers food becomes part of processes of struggle and resistance and redefinitions of home. Filipino domestic workers in Hong Kong found that the type of food they were expected to eat, the manner of eating (with chopsticks) and the small quantity of food provided alienating rather than comforting (Constable 2003).

Given the extent of control of domestic workplaces and leisure, domestic workers stretch home into the public sphere, recreating home beyond the nation and the house. Lisa Law (2001) writes evocatively of the 'little Manila' created in central Hong Kong every Sunday:

> In and around Central Hong Kong more than 100,000 Filipino women cast off the cultural conventions of their Chinese employers for one day a week, and eat Filipino food, read Filipino newspapers/magazines and consume products from an abundant number of Filipino specialty shops. ... The queues for the phone booths and at the post office are a sight in themselves, and when domestic workers are not posting letters or chatting long-distance they are writing and reading letters to and from distant loved ones. ... When experienced from the inside, however, Statue Square and its environs become a home from home for migrant women, a place of remembering and forgetting, and a lively place full of laughter, songs and home cooking.
>
> (266)

This sense of home, is, moreover, deeply connected to the senses of smell and taste:

> The Philippines is experienced each Sunday through a conscious invention of home – an imagining of place through food and other sensory practices that embody Filipino women as national subjects. 'Home cooking' thus becomes an active creation: a dislocation of place, a transformation of Central, a sense of home. In savouring Filipino food, for example, senses of taste, touch, vision and smell all become active. ... Putting aside chopsticks not only disrupts the routine of everyday eating, it is about enjoying eating with the hands. It is less about a refusal to use chopsticks than about enhancing flavour – a practice well known at home. The senses thus connect Hong Kong to the Philippines.
>
> (276)

As depicted visually in Figure 5.5, a sense of national belonging, of home, is hence constructed through 'home cooking' in spaces beyond those of the dwelling.

Figure 5.5 Filipino domestic workers in central Hong Kong on a Sunday. Reproduced courtesy of Lisa Law.

One of the most striking characteristics of contemporary domestic work is that it also extends the spatial extent of households and families. Many domestic workers leave husbands and children in their countries of origin, embarking upon 'diasporic journeys' (Yeoh and Huang 2000) in order to provide income for families 'back home'. As a result, home and its attendant mothering identities, are recast: '[f]ar from eroding notions of family ties, diasporic existence often serves to strengthen women's gendered identifications as sacrificial sisters, daughters, mothers, and wives' (Yeoh and Huang 2000: 422). Domestic workers in Singapore speak of their employment and location in terms of making sacrifices for their children. Moreover, home remains identified with the country in which they born, and is constructed as a place of eventual return. Phone calls and letters connect the women with 'home', but their relationship with this home is constantly renegotiated. In some cases, home becomes a fraught place. Letters, visits and phone calls home may become traumatic: 'I think I'm scared to go home, I dare not. It will only make me more homesick' and ' I usually cry while writing letters to them [children] and I cry whenever I receive letters from them' (domestic workers quoted in Yeoh and Huang 2000: 423; also see Asis *et al.* 2004; Mack 2004). Transnational domestic work thus stretches family homes across nations.

It is not only domestic workers for whom migration necessitates a transnational experience and imaginary of home. New, global, domestic spaces and spatialities of home have also resulted from the global mobility of professional workers and business people. Temporary periods of work in other countries have become increasingly prevalent for professional workers, with implications for the spatialities of the household and constructs of home. According to Hardhill,

> The term household is in fact usually associated with a locus of residence, but for the transnational households described in this paper there are two places of residence for the one household in two different countries.
>
> (2004: 386)

Transnational households hence create home that is, by necessity, connected to dwelling in neither the country of origin nor in one place. In some places, expatriate communities are constructed to embody transnational materialities and imaginaries of home. Skilled British workers living in France, for example, became 'immersed in translocalities within the host country such as socialising in British pubs … [and] living in a cultural-linguistic British domestic milieu' (Scott 2004: 393; see Research Box 9 by Katie Walsh). Home-based objects and practices were used to create a 'comfortable' English home:

> Expatriate homes were often adorned with vast video and book collections: purchased from specialist shops in Paris such as WH Smith's, supplied through internet stores like Amazon, sent by friends, or acquired during visits back to the UK.
>
> (Scott 2004: 402)

Like the family spaces constructed by transnational domestic workers, transnational professional workers' households also straddled two homes. Commuter households in which partners lived in different countries, or parents in one country and children in another, are one form (see Hardhill 2004). Considerable geographical attention has been paid to the phenomenon of 'astronaut families' and 'parachute children' for their apt illustration of the spatial stretching of home and familial displacement (Hardhill 2004; Ho 2002). These spatialities of family and home are most commonly associated with the global dispersal of wealthy Chinese, principally from Hong

Kong, in the 1980s and 1990s. Either through intent or circumstance, migration strategies took either an 'astronaut' or 'parachute' form. In the former, women and children would settle in the new country (commonly Canada, Australia or the United States), whilst the male partner would remain in Hong Kong in order to maintain business interests. Family relations would be maintained through the male flying across the world periodically (hence the term astronaut). In the latter case, the perceived cultural and economic opportunities provided by a transnational education would see children being 'parachuted' into new cities, where they would settle without their parents. Johanna Waters (2002) outlines how women living in astronaut families in Vancouver experience multi-locational homes. Whilst in one sense home was transnational for these women, its localization was intensified through the migration process. In the absence of paid domestic help and extended family, and in a new environment, traditional gendered roles were enforced. Women in astronaut families assumed a greater share of domestic and parenting responsibilities and were less mobile within cities because of fears of driving (Waters 2002: 120). Similarly displaced yet emplaced experiences of Vancouver as home emerged from David Ley's interviews with male Hong Kong migrants to Canada. Yet for these men, entering under a business migration programme, identities in their new homes were fractured by their sometimes failed business experiences. Gendered imaginaries and experiences of home are thus reinforced at the transnational scale.

MATERIALIZING HOME TRANSNATIONALLY: DOMESTIC ARCHITECTURE AND WESTERN IDEALS OF HOME

In Chapter 3 we discussed the importance of context in the numerous material forms (and in particular house design) taken by home. In the main, context has been defined as local: cultural ideals and gendered norms are expressed in house design, and the economic institutions producing housing (for example building and development firms) are predominantly local. However, in Chapter 4 we showed instances of transnational influences on, and the built form of, home, for example through the spatial mobility of the bungalow as an idea and built form of home. It is similarly instructive to think of high-rises as transnational homes in multi-layered ways. Like the bungalow, they are a globally dispersed material form of home, being a common housing form in nations as diverse as Singapore, the United States, the Philippines, Germany and Argentina. In some respects, then, the high-rise symbolizes a global domestic, a material form of home found across the world. But their transnationality also lies in their constructions of home within and across nations. Jane Jacobs (2001) uses the term 'hybrid highrises', drawing our attention to the ways in which the high-rise is domesticated and imagined in different national and historical contexts (see Jacobs 2006). In this section we wish to render this argument even more complex through a consideration of the transnational flows – economic, cultural and social – that constitute dwellings as home in the modern period. We begin with a discussion of the embeddedness of certain forms of housing, in certain locations, in global investment flows of transnational corporations. We then turn to the global imaginaries through which home is advertised as well as transformed.

Transnational housing

Housing is increasingly embedded in production and financial processes that stretch across the world. Whilst localized and national production systems remain important in various parts of the world (see Buzzelli 2004), globally extensive residential property markets and globally operating residential property corporations are assuming a greater prominence (Beauregard and Haila 1997). The price, design and availability of housing are hence connected to global financial systems and activities in other places. This is especially so in large-scale redevelopment projects such as Battery Park City in New York, London's Docklands (see Fainstein 1995), and what Kris Olds terms the Urban Mega Projects (UMPs) of the Pacific Rim (Olds 2001). UMPs can be found in a number of Pacific Rim cities, including Sydney, San Francisco, Jakarta, Singapore, Manila and Vancouver (Olds 2001: 31). In these places (see Figure 5.6) we see a growing homogeneity of house forms and contexts: collections of apartments envisaged largely for the affluent middle class, and located on previously industrial sites in the inner cities of large cities. These are transnational homes, then, in their growing transnational similarities.

These homes are also transnational in that the development of these new urban spaces, and houses within them, are strongly linked to globalization processes. Kris Olds nicely summarizes that these new urban spaces are 'modeled on each other', developed and planned by those with experience working on similar projects in other parts of the world, 'marketed to overseas firms and high income individuals' and 'designed to symbolize a global urban "utopia"

Figure 5.6 Housing on the North Shore of False Creek, Vancouver. Reproduced courtesy of Kris Olds.

for the twenty-first century' (Olds 2001: 6). Moreover, these developments embody and express transnational connections. The False Creek development in Vancouver, for example, was undertaken by the Concord Pacific Corporation and its owner Li Ka-shing. According to Olds, Li Ka-shing had strong social and business connections between his home in Hong Kong and Vancouver, including locating his son in Vancouver to gain an education and oversee the development. Apartments in False Creek were marketed in Vancouver and Hong Kong simultaneously and in similar ways through sales centres that produced images of living in these sites. Indeed, in some cases the Vancouver apartments were available for sale in Hong Kong before they were available in Vancouver (Olds 2001). Finally, these apartments embodied transnational notions of home in their openness to the world. The Concordia, for example:

> is no pedestrian condominium unit, for it is connected with 'state of the art fiber optics providing enhanced personal communications' enabling you to carry out a multitude of activities such as: working at home through the exchange of high speed text, audio, video and graphic

information with people from around the world; calling home to turn on the oven or adjust room temperature; ordering videos to be 'piped in'; and, surveillance through the use of high definition video monitoring equipment to simultaneously view up to four different locations (by splitting your television screen into quadrants) in the private, semi-public, and public spaces in and around *The Concordia*.

(Olds 2001: 2)

Apartments such as these are far from unique, but do nicely illustrate the transnational openness of home in the contemporary period.

It is not only apartments in large-scale redevelopment processes that are explicitly about providing housing for a globally mobile population. As a 'world city', London has become home to many transnational professionals. Paul White and Louise Hurdley (2003) focus on Japanese professionals who relocated temporarily to London for work. In it they found that a distinct housing sub-market had developed to provide housing for this group. Japanese estate agencies were important cultural and economic intermediaries, providing a source of houses presumed to be appropriate to Japanese ideals of home yet located in London. With names such as 'London-Tokyo Property Services' (White and Hurdley 2003: 693) they help companies identify appropriate locations for Japanese homes in London (typically near Japanese schools or other Japanese migrants). They also work with (and perhaps construct) what are presumed to be the attributes of home desired by professional movers: 'Cleanliness, the state of the kitchen and the bathroom and, in particular, a plentiful supply of hot water' (696). From a housing studies perspective White and Hurdley point out these estate agents are important facilitators of transnational relocation. But in terms of transnational notions of home we can also say that these agents filter and reconstruct ideas of home: taking knowledge about what a Japanese professional wants in a home, and reconstructing these ideas about home in light of London housing.

The perspective of a critical geography of home might lead us to question whether these estate agents reproduce fixed and stereotypical notions of home in Japan, and the ways in which they do so. Students who move countries for higher education are another important group creating transnational networks and homes, particularly in countries such as Australia where education is a growing export industry. Ruth Fincher and Lauren Costello (2003) highlight how understandings of ethnicity permeate the thinking of those responsible for providing housing for these transnational students. In some cases universities, in consort with property developers, build apartment blocks for international student housing. Yet these housing providers presume that they need to provide 'housing for wealthy consumers, accustomed to high rise and high cost dwellings' (Fincher and Costello 2003: 170). Home-making practices that will be accommodated in small, private apartments are assumed. One manager of a high-rise student housing development said:

> I believe and this is my personal opinion, and I would say most of our crew agrees with it, that the international market is more suited to this style of accommodation, and I am talking about Southeast Asian.
>
> (quoted in Fincher and Costello 2003:171)

Rather than challenging or opening up ethnicity-based notions of home, the providers of housing for transnational students reinforce them. This is underscored by the experiences of transnational students in more conventional residential colleges in Melbourne. Here, the domestic practices of these students are seen as non-conforming, and 'not fitting in'. Being at home in the residential college environment entails 'adherence to a collective spirit, [and] a willingness to share one's living space with others' (Fincher and Costello 2003: 181). Yet there is worry from college managers that this will not occur. According to one:

> I've appointed this year some International Student Liaison Officers [from] the student body and try and involve them. And it's very difficult. But they get, of course, involved in their own subgroups along ethnic or national lines. And they seem to be happy enough with that, some of them.
>
> (181)

Transnational homes, in the case of university students, disrupt, yet are interpreted through, dominant ideals of home.

Transnational imaginings of residence

In Chapter 4 we highlighted the ways both the built form and imaginary of the bungalow 'travelled' with, and were modified through, imperial encounters. In the twenty-first century, we can make a similar argument concerning the connection between suburban houses and neighbourhoods and globalization processes. The suburban house and neighbourhood have become even more prolific transnational housing forms. In China, for example, Fulong Wu (2004) alerts us to the transplanting of western 'city-scapes' in new urban developments in Beijing. He describes a new residential estate in Beijing called 'Orange County', marketed to the affluent middle class, in the following way:

> The project boasts '100 percent authentic design' drawn from the same project in the USA. The original design, according to widely distributed brochures, won a prize for New Homes in the USA in 1999. Under the title of 'Beijing's pure European American villa', another source emphasizes its authenticity: 'adopting the original American style, the Orange County project uses a blueprint that won the 1999 California Gold Medal'.
>
> (Wu 2004: 227)

Whilst the developers and builders do not operate globally, they do produce a vision of the domestic relying upon transnational imaginaries:

> To boost authenticity, developers adopt various innovative measures including employing global architects, mimicking Western design motifs, naming the roads and buildings with famous foreign names that are familiar to the Chinese, and even forging a relationship of sister communities with foreign towns.
>
> (Wu 2004: 232)

This global domestic is, nonetheless, also reliant upon, and modified through, local domestics. In the case of Orange County and similar developments in Beijing, property developers pursue their global strategy in order to capture and further solidify niche residential markets that have emerged in Beijing. In particular, recent marketization processes have seen enlarged residential stratification in the city, and these globally-informed housing imaginaries build upon and foster this stratification.

Sitting alongside transnational house styles such as these Beijing townhouses are 'transplanted' neighbourhoods, which become locations of transnational homes. The gated community is perhaps the most well-known transnational neighbourhood as home. First identified in the United States characterized by walls surrounding houses, limited entry and surveillance of residents and visitors, gated communities are now also present in countries across Europe, Asia and other parts of the southern hemisphere (see Atkinson and Blandy 2005). Again, the case of contemporary China (though see also Leisch 2002 for examples from Indonesia) succinctly illustrates how housing and imaginaries of home become connected to spaces across and in between nations. In some instances, these neighbourhoods literally become transnational homes, functioning as expatriate enclaves and home to the increasingly large number of foreigners employed by multinationals in China (Wu and Webber 2004). In the alternative instance of 'commodity housing enclaves', the American-inspired gated community travels to create luxury housing developments for native Chinese in Beijing, Shanghai and Nanjing (Wu 2005). In these places, home becomes a very private, secluded place as well as a place to be anonymous and feel secure. These homes are socially exclusionary, dependent on distancing and separation from marginalized others.

Notions of home in gated communities such as these in China remind us of the possibilities in thinking about the ways in which the ideal of home, as well as its manifestation in the suburban house form, are transnational (see also Box 5.7). The social construct of the ideal home was historically embedded in the unique social, economic and cultural circumstances of the United States, Britain, Canada, and Australia, as we explored in Chapter 3. Though there have always been transnational elements in this ideal, transnational flows of the ideal of home are now more obvious. One interesting example is provided by Oncu's research in Istanbul, Turkey (Oncu 1997). During the 1980s and 1990s, home for Istanbul's middle classes was redefined through a combination of exposure to popular culture narratives and discourses surrounding new housing. In both these narratives home as a place separate from the city was emphasized:

> a lifestyle cleansed of urban clutter – of poverty, of immigrants, of elbowing crowds, dirt and traffic – a world of safe and antiseptic social spaces where the 'ideal home' signifies clean air, clean water, healthy lives; a homogeneous setting and a cultural milieu where adults and children lead active lives, engage in sports, socialize with each other around their barbecue sets in the gardens.
>
> (Oncu 1997: 61)

This ideal was also dependent on a nostalgia for homes of the past, remembrances of an Istanbul full of unspoilt natural beauties (Oncu 1997: 63). This ideal home takes a number of material forms: garden cities as well as high-rise apartments. In the latter, termed 'sites', the modernity of home, and home as an opportunity for consumption becomes paramount:

> refrigerators, washing machines and kitchenware require 'modern' bath-rooms and kitchens; matching living-room furniture and television sets demand rooms to display them in.
>
> (Oncu 1997: 68)

Ideals of home are thus transnational, refracted through the political and social geographies of the local and the global.

The high-rise brings us to our final sense of the ways in which the domestic is being used to forge a national identity in a transnational world.

Box 5.7   TRANSNATIONAL IMAGININGS OF THE SUBURBAN HOUSE-AS-HOME

Anthony King has alerted us to the global dispersal and modification of the bungalow as a house form and we have suggested in this chapter that the late twentieth century saw the firming and extension of this process. Transnational migration and intensified and more frequent transnational movements of capital and images have also ushered in changes in the ways in which the suburban house-as-home is imagined. Imagery used to represent the suburban house has long been driven by context. In Australia, for example, Kathleen Mee (1993) has demonstrated strong representational links between ideal homes and other suburbs of Sydney. Rather than abstract refer-ences to power and heritage, for example, housing consumers are presented with references to powerful and history-laden suburbs of Sydney. Imaginaries of home are dependent on, and reproduce, geographies. Kim Dovey (1999) has more recently conducted a similar analysis at a transnational scale and provides an interesting illustration of the ways the domestic dreamings of the suburban house are expressed transnationally. He presents and interprets the names of houses in advertising brochures in both the United States and Australia, and finds striking similarities. Dovey's (149) analysis is depicted in a simplified form in Figure 5.7.

The ideal home is:

- found in other places and times;
- enveloped in nature;
- securely based in an unchanging past; and
- linked to power and success, though often through nature, heritage and place.

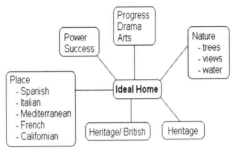

Figure 5.7   The ideal home (based on Dovey 1999: 149).

In Singapore, for example, the high-rise was part of the Singaporean government's attempts to modernize Singapore, to symbolize its modernity and to erase racial divisions (see Figure 5.6). Thus scientifically rational, and the most up-to-date, building technologies and forms were used in building high-rises. One of the intents of the similarity of built forms, including identical apartment layouts, was to symbolize a lack of racial difference in the Singaporean nation (Goh 2001:1590). The increasingly transnational orientation of the Singaporean economy and politics has seen a subtle shift in the role of the domestic in nation-building projects. New and old homes have been undergoing 'stylistic makeovers', partly as a way of cementing Singapore's global position. A 1999 speech of the Prime Minister was entitled 'First-world economy, world-class home'. The appropriation of transnational design motifs – 'English tudor, art deco, mission, Chinese kitsch' (Goh 2001: 1597; also see Jacobs 2006) – is seen to construct such a 'world-class' home, combined with increased recognition of capital gain from housing. Home is a transnational space in Singapore – through the involvement of transnational architects, builders and imaginaries. The transnationalism of Singapore is also facilitated and imagined through its

Figure 5.8   Recent high-rise estate in Woodlands, Singapore. Reproduced courtesy of Jane M. Jacobs.

homes. Anthony King and Abidin Kusno (2000) highlight a similar role for transnational homes in China. The architectural and urban design practices of villas and high-rise apartments 'are being staged in order to establish the central position of China vis-à-vis its Asian counterparts as well as the world at large' (King and Kusno 2000: 42).

CONCLUSIONS

This chapter has canvassed the numerous ways in which transnational mobility rests upon, yet reconfigures, material and imaginative geographies of home. We have considered the migration of diverse sets of people such as those who voluntarily and involuntarily migrate across national borders (and sometimes back again) as well as their children. We have also considered the transnational stretching of both ideals of home and the economic organizations that produce dwellings. Across these diverse fields the chapter has demonstrated a number of the key elements of a critical geography of home. First, it has shown how imaginative geographies of home involve attachments to more than one place and that such diffuse geographies of home are materially created and maintained (for example through sending remittances or materializing homeland in dwellings). Second, the links between power, identity and home have been clearly evident, especially in differences between those who migrate voluntarily and involuntarily, the working conditions of many transnational domestic workers, and experiences of dispossession. Finally, this chapter has explicitly addressed the multiple scales at which home is constructed, with transnational communities shaped by the interplay of both mobile and located homes and identities and by the processes and practices of home-making both within particular places and across transnational space. We continue this focus on multiple scales in the next and concluding chapter where we not only summarize the key arguments of the book but also bring together the different elements of a critical geography of home in two case studies.

Research box 9   EXPATRIATE BELONGINGS

Katie Walsh

My PhD research is an ethnography of British people living in Dubai, through which I explore expatriate belonging by focusing on embodied geographies of foreignness, material geographies of domesticity and emotional geographies of intimacy. Notions of home, as both an imagined sense of belonging and a lived space, are central to my research in several ways.

British expatriates in Dubai are an example of a transnational or diasporic group, so their experience of home highlights the multiplicity and fluidity of home more generally. To understand some of this complexity, I conducted participant-observation inside expatriate homes and in-depth interviews with householders. Talking about people's *belongings* (in the sense of possessions) and some of the ways that expatriates used domestic material culture in their daily lives, enabled me to think about the interdependence of imagined and material diasporic homing. Some objects, whether banal or extraordinary, become particularly significant in practices of remembering, something that became evident in the emotional stories about the process of sorting, choosing, packing and moving things to Dubai, as well as in narratives surrounding life-samples. Likewise, home-based leisure activities or work – such as arranging the furniture, cleaning, watching a DVD or shopping – are domestic practices that continually reiterate the significance of the lived space of home in less-tangible notions of belonging. Irrespective of whether an expatriate's experience of moving involves a whole house full of possessions or a single memento, it seems that they are attached to the stuff that makes their home special.

Although other chapters in my thesis rarely discuss 'home' explicitly, the language of belonging they employ resonates with wider research on home, reflecting the way these terms are often used synonymously. I analysed British expatriates' experience of, and interaction with, 'foreignness', focusing on some of the ways in which a national sense of home is performed in everyday racisms abroad, as well as how the unsettling experience of 'culture shock' is related to a feeling of being away from a homeland. My research also explored the significance of textures of expatriate intimacy to notions of belonging. The emotional geographies of a range of interpersonal relationships – including friendship, kinship, marriage and dating – suggest the centrality of intimacy in notions of being at home or away from home, and in (gendered) experiences of making homes abroad. Finally, my thesis also raises questions about home during more personal reflection on my own experiences of field work. To carry out ethnographic research it was important that I lived in Dubai, so I became a British expatriate for seventeen months and made it my temporary home. Disrupting my own emotional geographies in this way enabled me to think about the limitations of opposing notions of being home and away and its use in supporting conventional ethnographic authority.

Clearly, therefore, for British expatriates in Dubai 'home' is a central concept through which belonging is grounded, understood and negotiated in everyday life. This makes analysing notions of home and home-making practices a vital part of research on migrant and mobile identities in the contemporary world.

Katie Walsh is completing an ESRC Postdoctoral Research Fellowship in the Geography Department, Royal Holloway, University of London. In 2005 she completed her PhD, also at Royal Holloway, with the title 'British Expatriate Belonging in Dubai: Foreignness,

Domesticity, Intimacy' (ESRC postgraduate training award R42200134499). Katie's ethnographic research focuses on the significance of material culture, relationships and the body in everyday practices of belonging for the British abroad.

Research box 10   HOME AND MIGRATION IN JAMAICA KINCAID'S NOVELS

Rachel Hughes

I was not especially interested in 'home' when I began reading the novels of Jamaica Kincaid in the course of researching geographies of empire. I had been interested in colonial dominions, the transaction of cultural forms and people between places within the (former) British Empire, and in what it was that literature, specifically women's writing, could bring to geographical theory. However, reading Kincaid sparked my interest in the problems of home.

Kincaid was born in 1949 on the British dominion of Antigua in the Eastern Caribbean, and migrated to New York City at age sixteen. I first read *Annie John* (1986), Kincaid's semi-autobiographical novel about a childhood in Antigua. I then read the novel *Lucy* (1990), also narrated by a young Afro-Caribbean woman, about the experience of migrating from the Caribbean to a large city in the United States. I soon realized that to understand Kincaid's imaginative geographies of Antigua (as a colonial dominion), and the migration experience (in a post-imperial world), I would have to attend to domestic space and its constitutive experiences and relationships.

The words 'domestic' and 'dominion' share an etymological root: dominion is from *dominus*, lord of the *domus*, the home. Anne McClintock, writing of this shared etymology, observes that 'domesticity denotes both a *space* (a geographic and architectural alignment) and a *social relation to power*' (McClintock 1995: 34). Kincaid has stated in an interview that she is a writer 'who is very interested in domestic life' (Kincaid quoted in Perry 1983: 136–7).

In *Annie John* (1986), a myriad of domestic objects are 'Made in England'. Ceramic washbasins, bookshelves, felt hats and English herbs occupy 'pride of place' in Kincaid's lower middle-class home of 1950s and 1960s Antigua. In *At the Bottom of the River* (1978) Kincaid writes of glasses on a shelf that 'commemorate a [British] coronation' (13). In *The Autobiography of My Mother*, the character Ma Eunice treasures a plate that depicts 'a wide open field filled with grass and flowers in the most tender shades of yellow, pink blue and green', about which the narrator states: '[the plate] had an atmosphere of secret abundance, happiness and tranquillity' (9). The 'naturalised' deference to the signifying power of the (British) Motherland is paralleled in the expectation that the child will defer to the mother (and father) within the home. The cooperative and loving domesticity of early childhood, in which mother and daughter live in mimetic intimacy, is soon ruptured by the maturing child's refusal to be placed within the social relations of imperial power. Narratives of obedience, provision of food and care for the father, and concern with sanitation and feminine grooming require that the mother labour tirelessly and repress her feminine sensuality. The child comes to disrupt these domestic roles and expectations in a variety of ways, including fantasies of matricide and (spatialized) disobedience, such as the following:

> I would quietly unlatch the gate; creep back into the yard, and dive under the house to extract or hide some object that was forbidden me – usually some object that had come into my possession through my expert stealing.
>
> (*Annie John*, 66)

Below-house, the street and the classroom are all spaces that afford pleasure and power to the rebellious child. In the novel *Lucy*, however, the refusal of home spurs a young woman's emigration. While the narrator, Lucy, experiences new domestic restrictions and indignities in her work and lodging with a wealthy, white American family, she does not wish to return 'home'.

The British-Caribbean homes of Kincaid's writing directly challenge the pre-eminence of spaces of exploration, adventure, education and governance in both colonial and postcolonial cultural representations, foregrounding instead the power of the mother/ Motherland in the home. Protagonists repeatedly refuse a nostalgic relation to 'home' while being 'away'. Kincaid's homes are significant to theorizations of home more broadly because they attest to disempowerment along axes of race and gender in familial relationships under imperialism, associating profound ambivalence and destructive intimacy with being 'at home'.

Rachel Hughes is a lecturer in geography at the University of Melbourne. Her research interests encompass questions of mobility, memory, critical geopolitics and visual and material cultures.

6

LEAVING HOME

At the beginning of this book we asked 'what does home mean to you?' To what extent, since reading the book, has your answer to this question changed? How do your own experiences and feelings about home relate to the critical geography of home that we have developed in this book? Hopefully you have begun to think more critically about the ways in which the home is shaped by a wide range of socio-political and cultural processes that vary over time and space, whereby what home is, and what home means, depends to a large extent on who you are and where you are. We hope that you have thought about familiar, homely, and often taken-for-granted aspects of home in more critical ways, and that you have also learnt about less familiar, unhomely and more unsettling notions of home.

This concluding chapter has two parts. First, we return to the conception of home that we introduced in Chapter 1 and have developed throughout the book. Our intent in this section is both to refresh your understanding of home and to draw out connections between the chapters. Second, we deploy a critical geography of home to analyse two contrasting examples of home in the contemporary world: the destruction and reconstruction of home as a result of disasters, particularly Hurricane Katrina in the southern United States in 2005; and new visions of home that are materialized through cohousing. We discuss these two examples to demonstrate the ways in which a critical geography of home can be employed to understand a wide range of home-related issues that are important for understanding the contemporary world.

REVISITING HOME

Throughout the book, we have argued that the home should be understood in both spatial and political terms. Understanding home as both spatial and political underpins what we have termed a critical geography of home, which has three key, overlapping, components: home as material and imaginative; the nexus between home, power and identity; and the multi-scalarity of home.

Home as material and imaginative

Home is much more than a house or the physical structure in which we dwell. Home is both a place or physical location and a set of feelings. The argument of this book has been that home is a relation between material and imaginative realms and processes, whereby physical location and materiality, feelings and ideas, are bound together and influence each other, rather than separate and distinct. Moreover, home is a process of creating and understanding forms of dwelling and belonging. Home is lived as well as imagined. What home means and how it is materially manifest are continually created and re-created through everyday home-making practices, which are themselves tied to spatial imaginaries of home.

One theme that has run throughout the book and clearly illustrates the material and imaginative geographies of home is domestic architecture and design. Simultaneously material and imaginative, domestic architecture and design produce built forms, domestic interiors, and material objects which are themselves shaped by, and interpreted through, a wide range of ideas about home. In Chapter 2, for example, we discussed various ways in which the built domestic form has been mobilized by artists, writers and museum curators to convey and inspire a wide range of meanings, experiences and emotions about home. At the same time, everyday home-making practices, such as the display of family photographs and other objects within the home, are creative and imaginative as well as materially grounded, binding lived experiences of home to memories of past homes, which often stretch over transnational space. In Chapter 3, we explored the ways in which the suburban family home reflects and reproduces normative understandings of home and family. We critiqued the ways in which such domestic forms and families are represented as 'ideal', homely homes, and traced a range of ways in which the material spaces of home are contested and recast to reflect different, more inclusive, imaginaries, lived experiences and relationships. Concentrating on different forms of domestic architecture such as the bungalow and the high-rise in

Chapters 4 and 5, we explored the ways in which ideas about home take material form over imperial, national and transnational space, and become sites for the renegotiation of home as a place of identity and belonging.

The relational geographies of both material and imaginative homes are part of our broader concern to unsettle a series of dualisms or oppositions that have been influential in thinking about home. For example, home has been understood to be part of the private rather than the public spheres, a space for women rather than men, and bound to the spheres of social reproduction and the scale of the house rather than production and the global scale (see Box 1.2). Whilst in Chapter 1 we developed a theoretical critique of such dualistic thinking about home, the examples and analysis in the subsequent chapters present a critical geography of home that foregrounds simultaneity and connections rather than oppositions. Further, we have emphasized the importance of exploring the construction of these elements and their interconnections through political, social, economic and cultural processes. Some of these connections are listed in Table 6.1.

Throughout the book there are numerous instances of these connections and their embeddedness in social processes. Chapter 3 may have been organized as a discussion of homely and unhomely homes, but also demonstrated the unhomeliness of homely homes and vice versa. The simultaneous public and private nature of house-as-home was emphasized in the ways that familial and intimate experiences of dwelling are always infused with direct or indirect actions of the state. Chapter 4 illustrates the porosity of home through the display of cosmopolitan objects, but also foregrounds attempts to place boundaries around the nation-as-home, especially in the contemporary politics of homeland security. In Chapter 5 we showed how transnational

Table 6.1 Home as relational

Homely	and	Unhomely
Public	and	Private
Local	and	Global
Material	and	Imaginative
Masculine	and	Feminine
Bounded	and	Porous
Located	and	Mobile
Nature	and	Culture
Site of alienation	and	Site of belonging

migration reconfigures both material and imagined elements of home and the links between them, as shown by the emotional and material connections between people, homes and homelands across diaspora space. Finally, we have been careful not to place an *a priori* positive evaluation on any of the terms used to constitute home, but have instead drawn attention to the ways in which such evaluations are constructed and embedded in relations of power.

Home, power and identity

Home as a place and as a spatial imaginary helps to constitute identity, whereby people's senses of themselves are related to and produced through lived and metaphorical experiences of home. These identities and homes are, in turn, produced and articulated through relations of power. In Chapter 3, for example, we explored the dominant meanings of home that include family, patriarchal gender relations, stability, security, and home ownership, whilst in Chapter 4 we focused on the importance of the home in forging and articulating the politics of nation and empire. Throughout the book we have interrogated rather than accepted such normative notions of home, investigating the ways in which they are not only mobilized in social processes, but are also resisted, reworked and transgressed by different social groups. So, for example, we have explored home-making practices by gay men and lesbians that resist the heteronormativity of home (Chapter 3), the political mobilization of the home in anti-imperial and indigenous politics (Chapter 4), and the reworking of home by transnational migrants (Chapter 5).

Two themes that have run throughout the book clearly illustrate the critical interplay of home, power and identity: *domestic work* and the relationships between *home and state*. Paid and unpaid domestic work is closely bound to power relations that exist beyond, as well as within, the household. The gendered nature of domestic work is closely tied to normative ideas about home and family, as we discussed in Chapter 3. In the following chapter, we showed how the employment of servants in British India underpinned the exercise of imperial power on a domestic scale by creating 'empires in the home', which were sites of imperial anxiety as well as dominance. Finally, in Chapter 5 we focused on the transnational migration of domestic workers, reflecting on the global inequalities that underpin this migration and the ways in which ties to home and family become stretched and recast over space. In each case, domestic work – and the ways in which it perpetuates gender, class and racialized inequalities – reveals the critical intersections of home and state, whether in terms of imperial expansion, social welfare policies, or state programmes that facilitate the migration of domestic workers. We have discussed the relationships between home and state in other contexts too, including the mobilization of home and homeland in contemporary security and welfare politics (including William Walters' ideas about 'domopolitics'), the fissures and new forms of belonging that are created when the home, rather than state-funded institutions, becomes a site of terminal caregiving, and debates over the rights of people – settler populations, indigenous people, refugees and asylum seekers, for example – to regard the nation-as-home (see Box 6.1 for further research ideas on this and other areas).

Home as multi-scalar

As we have shown in relation to the material and imaginative geographies of home, and the relationships between home, power and identity, a critical geography of home recognizes its multi-scalarity. Our argument about scale has been twofold. First, we have argued that home is a socially constructed scale that extends beyond the house and household. Thus for suburban families, house is home, but home-making practices also extend to include the wider suburb or neighbourhood. Many British women living in imperial India sought to recreate familiar homes in their Indian bungalows and in hill stations, but found it profoundly unsettling to return 'home' to Britain. As we discussed in Chapter 5, diasporic home-making fosters and recasts connections over transnational space through, for example, domestic architecture, particular objects within the home, and through the domestic preparation and consumption of food. Second, we suggest that the relations of domesticity, intimacy and belonging that construct home not only extend beyond, but also help to produce scales far beyond the household. Following Amy Kaplan (2002), for example, we explored the ways in which ideas about the 'domestic' are inseparably bound to ideas about the 'foreign', with material and imaginative implications for indigenous people who were displaced and dispossessed by imperial nation-building, and for those people who have experienced racial violence and feel increasingly insecure in the context of contemporary politics of homeland security.

The relationships between *home and family* illustrate the multi-scalarity of home. In Chapter 3, for example, we explored the ways in which homely and unhomely homes are bound up with notions of the family on a household scale. Moving beyond the household scale, in Chapter 4 we explored the ways in which conceptualizations of home and family have been mobilized to articulate and legitimate the nation and empire as home, binding the household and the domestic to the wider exercise of imperial power, anti-imperial nationalism, and the politics of the nation-as-home and homeland. We introduced the transnational mobility of home in Chapter 4 in relation to imperial resettlement, home-making and nation-building, and developed this in relation to contemporary migrations in Chapter 5. In this chapter, we explored the ways in which the relationships between home and family are reconfigured over transnational space, as shown by hybrid home-making across different generations living in diaspora, the transnational geographies of home for 'third culture kids', 'astronaut', and 'parachute' families, and the ways in which family relationships are recast in often very difficult ways for transnational domestic workers.

Throughout the book, we have argued that the home – as both lived experience and as spatial imaginary – is a site of violence and alienation as well as comfort and security. The multi-scalarity of home, and the relationships between home, power and identity, are clearly evident in our discussion of *violence* at home on various scales. Domestic violence, as we showed in Chapter 3, is experienced within the home, often remains 'private' and unreported, and is a major factor that leads women to become homeless. We also considered violence and the home in two different contexts in Chapter 4: first, the violent displacement and dispossession of indigenous people as a result of imperial nation-building; and, second, the racial violence suffered by those targeted as 'foreign', and thus out of place, in the national homeland. In Chapter 5 we considered the violence and abuse suffered by transnational domestic workers and the organizations that campaign for legal and human rights to protect such workers, who are particularly vulnerable as they often live as well as work in their employers' homes.

CRITICAL GEOGRAPHIES OF HOME

In the second part of this chapter, we deploy our argument about a critical geography of home to interpret two contrasting examples about home in the contemporary world (Box 6.1 suggests ways of approaching other issues raised by a critical geography of home). We begin by discussing the destruction and reconstruction of homes as a result of Hurricane Katrina, before turning to new visions of home materialized through cohousing. In both cases, we consider the ways in which the homely and the unhomely are not distinct and separate, but rather fold into one another in the constitution of home.

Hurricane Katrina

Throughout the book, we have considered a range of examples of the destruction and loss of home, and their material effects in terms of homelessness, dispossession, and forced migration. As we discussed in Box 4.6, Porteous and Smith (2001) describe the destruction of home in terms of 'domicide'. Whilst Porteous and Smith argue that domicide results directly from the agency of powerful people involved in corporate, political or bureaucratic projects, we focus here on the destruction of home as a result of a 'natural' disaster. In 2004–5, the large-scale destruction and loss of home has been both an immediate and longer-term consequence of the Asian tsunami, which devastated parts of Indonesia, Thailand, Sri Lanka and India in December 2004; Hurricane Katrina, which caused extensive damage to New Orleans in late August and early September 2005; and the major earthquake in northern Pakistan and Kashmir in October 2005. Numerous facets of these phenomena require analysis, ranging from assessments of the adequacy of official responses, understanding the scale and extent of personal and business philanthropy that followed many of the events, through to charting the links between these localized phenomena and global forces such as climate change. These events are also about home – disruptions to belongings and attachment as well as the loss of shelter, the porosity of home, and the formal and cultural politics of home. In this section we use just one of these examples to explore the contemporary purchase of a critical geography of home: the impact of Hurricane Katrina in New Orleans and the Gulf Coast of the United States in late August and early September 2005.

After travelling through the Gulf of Mexico, Hurricane Katrina hit land ninety kilometres south of New Orleans on 28 August 2005. Much of New Orleans lies below sea level, with the city protected by a series of levees. These levees were breached by the storm, and by 31 August an estimated 80 per cent of New Orleans was under water (http://news.bbc.co.uk/1/shared/spl/hi/americas/05/katrina/html/satellite.stm). Whilst some residents had left the city, many had not been evacuated when their houses and city become uninhabitable due to flooding. As a result, more than 900 people died, houses were destroyed or became uninhabitable, and people became homeless and were forced to flee their home neighbourhoods and city for elsewhere. According to an article in *USA Today* in October 2005,

> The American Red Cross estimates that more than 350,000 homes were destroyed by Hurricanes Katrina and Rita, while an additional 146,000 had major damage. Overall, 850,791 housing units were damaged, destroyed or left inaccessible because of Katrina. Looked at another way, the number of homes destroyed is equal to about 17 per cent of annual home construction, which is running at a pace of about 2 million a year.
> (http://usatoday.com, 6 October 2005)

What might a critical geography of home highlight in this situation?

First, a critical geography of home would highlight the multiple and overlapping scales through which the loss of home was understood and reactions to it. The Federal Emergency Management Agency (FEMA) – the agency responsible for coordinating the federal government response to disasters such as hurricanes, earthquakes and floods in the United States – comes under the jurisdiction of the Department of Homeland Security (discussed in Chapter 4). The role of FEMA indicates that the task of 'preparing America' to face emergencies and disasters lies on a domestic as well as a national scale: 'Educating America's families on how best to prepare their homes for a disaster and tips for citizens on how to respond in a crisis will be given special attention at DHS' (www.dhs.gov; also see the webpages of the American Red Cross, which include detailed instructions for preparing homes to face disasters and assessing the safety of a new home: www.redcross.org). Transnational and domestic scales were also linked through the quests of New Orleans evacuees for information on the homes and belongings they had left behind. Many turned to the internet, and used information proved by 'a grass roots effort that had identified scores of posthurricane images, determined the geographical coordinates and visual landmarks ... and posted them to a Google Earth bulletin board' (Hapner 2005: 1).

Second, Hurricane Katrina can be understood through the lens of strong emotional attachments to places constructed as home which may include, but are not confined to, physical shelter. Many people lost their homes and belongings. As well as reporting this loss of shelter, newspaper stories are replete with accounts of feelings of loss and aimless wanderings. Most profoundly, one woman who had been forced to flee told a reporter: '"When someone says the roots are gone, what happens to the rest of the plant?" Ms.

Allum said. "It dies. New Orleans is our roots"' (Zernike and Wilgoren 2005: 1). But we might also be alert to home-making practices in the context of loss: how and to what extent new homes are made that stretch across spaces and social divides. For example, the *New York Times* reported new ties forged between survivors, and how evacuees made their makeshift accommodation home (Medina 2005: 20).

Third, a critical geography of home could also highlight the influence of social divisions in the provision of house-as-home. The devastation caused by Hurricane Katrina exposed the poverty and inadequate housing of a disproportionate number of African Americans in the states along the Gulf of Mexico coast:

> nearly 60% of the counties declared disaster areas after Katrina – many with low-income African-American areas – already had poverty rates above 20% and large numbers of substandard houses. About 40,000 households in Louisiana, Mississippi and Alabama lacked adequate plumbing before the storms … About 50,000 households that had been receiving federal housing aid were displaced. The National Low Income Housing Coalition estimates more than half the housing destroyed by Katrina was rentals, and about 70% was affordable to low-income renters.
>
> (http://usatoday.com: 6 October 2005)

According to the Insurance Services Office, claims for property insurance after Hurricane Katrina are estimated at $34.4 billion (http://usatoday.com; http://news.independent.co.uk). Many other displaced people who were already living in poverty had no insurance for the homes and possessions that they lost. Also worth investigating is the role played by housing conditions and tenure in casualties from the hurricane, such as the deaths of 34 patients in a Louisiana nursing home (Rhode and Dewan 2005: 21). Knowledge of tenure divisions and social inequalities in relation to house-as-home could be used to evaluate government attempts to re-house people. Are, as some have suggested, policies to re-house people after disasters more applicable to owners rather than renters? (see *The Economist* (Anon 2005): 37).

Hurricane Katrina can also be approached through the lens of the domestic uncanny (see Kaika 2004). It overturned the constructed boundaries between 'natural' and 'domestic' spaces are that are both necessary and yet usually rendered invisible within the home. Home became unknown to itself, unhomely, because the natural was within, rather than outside it. A similar argument may be made in relation to safey. Just as terrorist and other threats reveal the ways in which the politics of homeland security in the United States are predicated upon, and also perpetuate, a politics of insecurity, Hurricane Katrina made visible the dangers and insecurities of home. Interpreting the effects of Hurricane Katrina in terms of a critical geography of home thus reveals the fragility, power and scale of home in the contemporary world.

Cohousing

Our second example is very different. As the work of Dolores Hayden (1996; 2002) and other feminists has shown, there is a long-standing tradition of architects, designers and social reformers who have sought to design and build non-sexist, inclusive homes. In large part, this tradition involves unsettling the normative assumption that an 'ideal' home is one inhabited by a nuclear family, and rather suggests ways in which people can live in more collaborative, collective or cooperative ways. Cohousing is one contemporary example whereby groups of people seek to design, build or adapt new visions of home (and part of the research discussed by Louise Crabtree in Research Box 5). The earliest cohousing project was built in Denmark in 1972. By 1993, more than 140 cohousing communities had been built in Denmark, ranging from six to forty households in size (McCamant and Durrett 1994). The term ' cohousing' was introduced to the English-speaking world in a book published by Kathryn McCamant and Charles Durrett of the American Cohousing Company in 1988 (visit www.cohousing.org; www.cohousingresources.com for more information). The first major cohousing community in the United States was completed at Muir Commons, Davis, California, in 1991 and, just two years later, more than 150 groups in North America were meeting to plan their cohousing communities (McCamant and Durrett 1994). Other cohousing projects have been built or are being planned in Europe and Australia.

What might a critical geography of home reveal about cohousing? First, cohousing offers an alternative vision of house-as-home in its relationships to neighbourhood and community through distinctive home-making practices. For McCamant and Durrett, for example, cohousing creates places 'that expand the meanings of "home," "neighborhood," and "community"' (1994: 6). Cohousing projects usually incorporate shared, collective space as well as private dwellings, balancing 'privacy and community' (Fromm 1991: 1). Rather than view the home as a private retreat from the world beyond, cohousing schemes provide collaborative alternatives for living. These alternatives are materially manifested in the built domestic form, as shown by the design of communal kitchens, gardens and other living space, alongside private houses and apartments. These alternative forms of home-making are also played out in the social interaction between residents that extends beyond the family through, for example, shared childcare, cooking, and care for older neighbours. As Dorit Fromm explains, in cohousing and other collaborative housing projects,

> each household has its own house or apartment and one share in the common facilities, which typically include a fully equipped kitchen, play areas, and meeting rooms. Residents share cooking, cleaning, and gardening on a rotating basis. By working together and combining their resources, collaborative housing residents can have the advantages of a private home and the convenience of shared services and amenities.
>
> (Fromm 1991: 7)

For many residents, cohousing creates new homes and home-making practices that are closely bound to new visions of neighbourhood and community. And yet, to what extent are these alternative forms of home-making exclusionary? Are cohousing schemes more similar to than different from the gated communities that we discussed in Chapter 5? As McCamant and Durrett explain, one concern, particularly in the United States, is that

> cohousing might further emphasize already existing American patterns of residential and social segregation. Certainly cohousing could be applied as just another variation on the walled-in, planned communities of the affluent; but such exclusivity runs counter to one of the primary reasons for the concept's appeal – the desire for an integrated residential environment. ... [Cohousing groups] make special efforts to integrate their community into the existing neighborhood. Cohousing offers an opportunity to overcome the current patterns of segregation by interest, age, income, race, and household composition that these people deem undesirable. In choosing cohousing, residents choose to respect each other's differences, while building on their commonalities.
>
> (202)

A second way that a critical geography of home can investigate cohousing lies in unsettling normative assumptions about home and family. Whilst many cohousing projects are designed to house families, and offer the possibilities of shared childcare beyond the nuclear family, others are designed to house residents in non-family groups. Some cohousing projects are cross-generational and house men and women, whilst others are designed by and for particular age groups, as shown by cohousing communities of older people, which are either mixed or specifically for women. In her study of cohousing for older people in the Netherlands, for example, Maria Brenton explains how this alternative vision of home provides many benefits compared to living alone:

> [Residents] have the quality of life that comes from remaining independent for longer in their own homes. They like belonging to a group they have chosen and they also like the privacy and freedom of having their own front doors. They like having people to do things with, the fact that there is always someone to talk to, the readiness of their neighbours to feed the cat and water the plants when they are away; they like the security of knowing that if they have a fall or get sick, their fellow group members will help them.
>
> (1998: vi)

This clearly illustrates the intersections of material and imaginative geographies of home, whereby the material design of the housing – 'having their own front doors', for example – coexists with a more collective, neighbourly vision of home as a place of companionship, security and care. This vision of home extends far beyond the physical structure of the dwelling, and does not necessarily depend upon family relationships. A number of cohousing projects are in development by and for older women, in part because of the disproportionate number of older women who live alone and live in poverty, including Older Women's Cohousing in London (OWCH; www.cohousing.co.uk/owch.htm).

Third, cohousing can be interpreted in relation to the multi-scalarity of home. In addition to viewing home in relation to family, neighbourhood and community, cohousing can also be analysed in terms of the critical geographies of home on national and transnational scales. The different ways in which cohousing can materially manifest new visions of home is closely tied to national housing markets and state housing policies. On a transnational scale, the different characteristics of cohousing schemes in, for example, Denmark, the United States and Britain not only reveal different geographies of housing, but also different geographies of home. Moreover, in some places, cohousing has become an important way for older migrants to feel at home. In the Netherlands, for example, in 1997 there were 25 cohousing communities and initiative groups for older people from immigrant communities, mainly from ex-colonies such as Surinam. As Maria Brenton explains,

> The CoHousing Community model can be a particularly appropriate one for older immigrants who may have a special need to substitute for disrupted traditions of family support and to create around themselves a linguistic and cultural milieu that is familiar and sympathetic. Frequently their financial position also presents difficulties if they have migrated late in life, have generally low levels of education, have been employed in low-paid work and have not built up supplementary pension entitlements. They need, in old age, to be located in the neighbourhoods where their families are concentrated and where an infrastructure of services such as ethnic shops and religious and cultural organizations can serve their needs. One group of Chinese elders in Rotterdam, for example, is situated in a building containing a Chinese cultural centre where meals and activities are already available.
>
> (Brenton 1998: 18)

In very different ways, the destruction of homes by Hurricane Katrina and the construction of alternative visions of home through cohousing show how a critical geography of home can be deployed to understand wide-ranging issues in the contemporary world. In both cases, and throughout the book, we have argued that the home is both material and imaginative, is situated within a nexus of power and identity, and is mobilized and recast over a wide range of scales. Rather than view the home as static, fixed and bounded, the home is in process, shaped by home-making practices, and embedded within wider social, political, cultural and economic relations on scales from the domestic to the global.

Box 6.1   STUDYING HOME: IDEAS FOR FURTHER RESEARCH

Many of the issues we have discussed in this book – such as home-land security, exclusionary housing provision, and belonging and alienation in an increasingly globally connected world – need much more geographical analysis. There are a number of other home-related contemporary issues that have not been subject to substantial geographical scrutiny, and we outline three of them here as points of departure from the book.

Home detention

Imprisonment at home is not new, but re-emerged in the 1980s in 'over 45 US states, the United Kingdom (UK), Europe, Asia, Australia and Canada' (Gibbs and King 2003: 1). In these countries the imprisonment of offenders at home rather than, or sometimes in conjunction with, detention in prisons, is related to ideological and financial constraints on government expenditures and changing philosophies of imprisonment and rehabilitation. In the United States, home detention, often accompanied by electronic monitoring of individuals, has increased since the mid-1990s, with one estimate being that in 2002 there were about 3,700 people in home detention across the nation (Kuczynski 2002). Most recently, for example, Martha Stewart – the American famous for her lucrative broadcasting and writing on home-making – was released from prison on condition of five months of home confinement (Hays 2005: C1). Outside the west home detention is more often used in the form of 'house arrest' of political dissidents. The long-term house arrest of Aung San Suu Kyi in Burma is most famous here, and a loose form of house arrest has also recently been used for Jiang Yanyong, the Chinese doctor who disclosed 'the true extent of Beijing's SARS epidemic' (Liu 2004: 49). Home detention raises a number of issues for geographers interested in home, including:

• How does home detention connect with, and disrupt, perceptions of nation-as-home?
• How is home imprisonment experienced?

- What identities are related to and transformed through home detention?
- What are the connections between home detention and various scales of state power?
- How do each of these vary with national and other contexts?

Implications of the Reconfiguration of State Power for Geographies of Home

Home detention is a specific instance of the wider phenomenon of the changing extent and workings of state powers. The relationships between home and state should be understood in terms of the multiple and contradictory workings and intent of state power. On the one hand there is a 'withdrawal' of the state, with functions previously undertaken by the state now being undertaken by private or non-government agencies. On the other hand, there has been an intensification of state involvement in everyday life, as shown by increased surveillance and monitoring of welfare recipients, new security policies, changing interventions in debates around, for example, abortion, palliative care, and euthanasia. The ways in which these contemporary reconfigurations of state power also reconfigure home are profound and hence worthy of further investigation. We have already drawn out the ways in which we have addressed the relationships between home and state in the book. Yet many questions remain, including:

- How are forms of belonging shaped by contemporary state power?
- What exclusions are involved?
- What presumptions of home are these new state policies dependent on?
- What, and how, is home experienced given its increasingly 'public' functions in relation to health, education and security?
- How might the concept of the uncanny help in understanding home in this new context?
- How are homes and belonging being transformed by neoliberal policies with respect to caring, imprisonment and education?

Nature, culture and the sustainability of home-making practices

We have touched on the extent to which home-making practices are socially and environmentally sustainable at various points in the book. For example, Box 3.7 outlined recent thinking on how to make housing more sustainable, whilst in this chapter we have outlined the consequences of 'natural' disasters for loss of home. Sustainable housing and home remains a pressing issue worth exploring. Geographers interested in home have much to contribute to this debate by addressing questions such as:

- What nature–culture relations are expressed in contemporary home-making practices and what are their implications for sustainability goals?
- What are the environmental consequences of home-making practices?
- How might ideals of home be refashioned in an environmentally sensitive way?
- How, if at all, can recognition of the multiple scales of home be used to open up debates on sustainability, home and housing?
- How might global events (such as world oil prices and climate change) influence home-making practices?

BIBLIOGRAPHY

Abramsson, M., Borgegard, L-E., Fransson, U. (2002) 'Housing careers: immigrants in local Swedish housing markets'. *Housing Studies* 17: 445–64.

Adamson, F. (2002) 'Mobilizing for the transformation of home: politicized identities and transnational practices'. In Al-Ali, N. and Koser, K. (eds), *New Approaches to Migration? Transnational Communities and the Transformation of Home*. London: Routledge, pp. 155–68.

Ahmad, M. (2002) 'Homeland insecurities: racial violence the day after September 11'. *Social Text* 72: 101–15.

Ahmed, S. (2000) *Strange Encounters: Embodied Others in Post-Coloniality*. London: Routledge.

Ahmed, S., Castañeda, C., Fortier, A.-M., and Sheller, M. (eds) (2003a) 'Introduction: uprootings/regroundings: questions of home and migration'. In Ahmed, S., Castañeda, C., Fortier, A.-M., and Sheller, M. (eds), *Uprootings/Regroundings: Questions of Home and Migration*. Oxford: Berg, pp. 1–19.

Ahmed, S., Castañeda, C., Fortier, A.-M., and Sheller, M. (eds) (2003b) *Uprootings/Regroundings: Questions of Home and Migration*. Oxford: Berg.

Ahrentzen, S. (1997) 'The meaning of home workplaces for women'. In Jones, J. P. *et al.* (eds), *Thresholds in Feminist Geography: Difference, Methodology, Representation*. Lanham: Rowman and Littlefield, pp. 77–92.

Al-Ali, N. and Koser, K. (eds) (2002) *New Approaches to Migration? Transnational Communities and the Transformation of Home*. London: Routledge.

Ali, S. (2003) *Mixed-Race, Post-Race: Gender, New Ethnicities and Cultural Practices*. Oxford: Berg.

Alibhai-Brown, Y. (1995) *No Place like Home*. London: Virago.

Allon, F. (2002) 'Translated spaces/translated identities: the production of place, culture and memory in an Australian suburb', *Journal of Australian Studies,* January: 101–10.

Allport, C. (1987) 'Castles of security: the New South Wales Housing Commission and home ownership 1941–61', in M. Kelly (ed.), *Sydney: City of Suburbs*, Kensington: University of New South Wales Press, pp. 95–124.

Anderson, K. (1999) 'Reflections on Redfern'. In Stratford, E. (ed.), *Australian Cultural Geography*. Melbourne: Oxford University Press, pp. 69–86.

Anderson, P., Carvalho, M., and Tolia-Kelly, D. (2001) 'Intimate distance: fantasy islands and English lakes'. *Ecumene* 8: 112–19.

Anon. (2005) 'A voucher for your thoughts: Katrina and public housing assistance'. *The Economist*, 24 September: 37.

Archer, J. (2005) *Architecture and Suburbia: From English Villa to American Dream House, 1690–2000*, Minneapolis and London: University of Minnesota Press.

Armbuster, H. (2002) 'Homes in crisis: Syrian Christians in Turkey and Germany'. In Al-Ali, N. and Koser, K. (eds), *New Approaches to Migration? Transnational Communities and the Transformation of Home*. London: Routledge, pp. 17–33.

Armstrong, N. (1987) *Desire and Domestic Fiction: a Political History of the Novel*. New York: Oxford University Press.

Arnott, J. (1994) 'Speak out, for example'. In Camper, C. (ed.), *Miscegenation Blues: Voices of Mixed Race Women*. Toronto: Sister Vision, pp. 264–8.

Asis, M. M. B., Huang, S., and Yeoh, B. S. A. (2004) 'When the light of the home is abroad: unskilled female migration and the Filipino family'. *Singapore Journal of Tropical Geography* 25: 198–215.

Atkinson, G. (1859) *'Curry and Rice' on Forty Plates; or, the Ingredients of Social Life at 'Our Station'*. London: John B. Day.

Atkinson, R. and Blandy, S. (2005) 'Introduction: international perspectives on the new enclavism and the rise of gated communities', *Housing Studies* 20(2): 177–86.

Attfield, J. (2000) *Wild Things: the Material Culture of Everyday Life*. Oxford: Berg.

Attfield, J. and Kirkham, P. (eds) (1995) *A View from the Interior: Women and Design*. London: Women's Press.

Australian Bureau of Statistics (2004) *Australian Social Facts*, Canberra: Australian Government.

Axel, B. K. (2001) *The Nation's Tortured Body: Violence, Representation, and the Formation of a Sikh 'Diaspora'*. Durham, NC: Duke University Press.

Bachelard, G. (1994) [1958] *The Poetics of Space*. Translated by M. Jolas. Boston, MA: Beacon Press.

Badcock, B. and Beer, A. (2000) *Home Truths: Property Ownership and Housing Wealth in Australia*, Melbourne: Melbourne University Press.

Ballhatchet, K. (1980) *Race, Sex and Class under the Raj: Imperial Attitudes and Policies and their Critics, 1793–1905*. London: Weidenfeld and Nicolson.

Barkham, P. (2002) 'Film forces Australia to face its past'. *Guardian Weekly*, 21 February (www.guardian.co.uk/GWeekly/Story/0,,654219,00.html).

Barr, P. (1976) *The Memsahibs: in Praise of the Women of Victorian India*. London: Penguin.

Barr, P. (1989) *The Dust in the Balance: British Women in India, 1905–1945*. London: Hamish Hamilton.

Bartrum, K. (1858) *A Widow's Reminiscences of the Siege of Lucknow*. London: James Nesbit.

Beauregard, R. A. and Haila, A. (1997) 'The unavoidable incompleteness of the city', *American Behavioral Scientist* 41(3): 327–41.

Beecher, C. (1841) *Treatise on Domestic Economy, for the Use of Young Ladies at Home, and at School*. Boston: Marsh, Capen, Lyon and Webb.

Beecher, C. and Stowe, H. B. (2002) [1869] *The American Woman's Home*. Edited by N. Tonkovich. London: Rutgers University Press.

Beetham, M. (1996) *A Magazine of her Own? Domesticity and Desire in the Woman's Magazine, 1800–1914*. London: Routledge.

Bell, D. and Valentine, G. (1997) *Consuming Geographies: We Are Where We Eat.* London and New York: Routledge.

Berthoud, R. and Gershuny, J. (eds) (2000) *Seven Years in the Lives of British Families: Evidence on the Dynamics of Social Change from the British Household Panel Survey.* Bristol: The Policy Press.

Bhabha, H. (1997) 'The world and the home'. In McClintock, A., Mufti, A., and Shohat, E. (eds), *Dangerous Liaisons: Gender, Nation, and Postcolonial Perspectives.* Minneapolis: University of Minnesota Press, pp. 445–55.

Bhatti, M. and Church, A. (2004) 'Home, the culture of nature and meanings of gardens in late modernity'. *Housing Studies* 19: 37–51.

Billig, M. (1995) *Banal Nationalism.* London: Sage.

Bird, C. (1998) *The Stolen Children: Their Stories.* Milsons Point, NSW: Random House Australia.

Bittman, M., Matheson, G., and Meagher, G. (1999) 'The changing boundary between home and market: Australian trends in outsourcing domestic labour'. *Work, Employment and Society* 13(2): 249–73.

Blickle, P. (2002) *Heimat: a Critical Theory of the German Idea of Homeland.* Rochester, NY: Camden House.

Bloch, A. (2002) *The Migration and Settlement of Refugees in Britain.* Basingstoke: Palgrave Macmillan.

Blodgett, H. (ed.) (1991) *'Capacious Hold-All': an Anthology of Englishwomen's Diary Writings.* Charlottesville: University of Virginia Press.

Blomley, N. (2003) 'Law, property, and the geography of violence: the frontier, the survey, and the grid'. *Annals of the Association of American Geographers* 93: 121–41.

Blomley, N. (2004a) 'The boundaries of property: lessons from Beatrix Potter'. *The Canadian Geographer* 48: 91–100.

Blomley, N. (2004b) 'Un-real estate: proprietary space and public gardening'. *Antipode* 36: 614–41.

Blomley, N. (2004c) *Unsettling the City: Urban Land and the Politics of Property.* New York: Routledge.

Blunt, A. (1994) *Travel, Gender, and Imperialism: Mary Kingsley and West Africa.* New York: Guilford.

Blunt, A. (1999) 'Imperial geographies of home: British domesticity in India, 1886–1925'. *Transactions* 24: 421–40.

Blunt, A. (2000a) 'Spatial stories under siege: British women writing from Lucknow in 1857'. *Gender, Place and Culture* 7: 229–46.

Blunt, A. (2000b) 'Embodying war: British women and domestic defilement in the Indian "Mutiny," 1857–8'. *Journal of Historical Geography* 26: 403–28.

Blunt, A. (2002) '"Land of our Mothers": home, identity and nationality for Anglo-Indians in British India'. *History Workshop Journal* 54: 49–72.

Blunt, A. (2003a) 'Home and identity: life stories in text and in person'. In Blunt, A., Gruffudd, P., May, J., Ogborn, M., and Pinder, D. (eds), *Cultural Geography in Practice.* London: Arnold, pp. 71–87.

Blunt, A. (2003b) 'Home and empire: photographs of British families in the *Lucknow Album*, 1856–7'. In Schwartz, J. M. and Ryan, J. R. (eds), *Picturing Place: Photography and the Geographical Imagination.* London: IB Tauris, pp. 243–60.

Blunt, A. (2003c) 'Collective memory and productive nostalgia: Anglo-Indian home-making at McCluskieganj'. *Environment and Planning D: Society and Space* 21: 717–38.

Blunt, A. (2005) *Domicile and Diaspora: Anglo-Indian Women and the Spatial Politics of Home.* Oxford: Blackwell.

Blunt, A. and Varley, A. (2004) 'Geographies of Home', *Cultural Geographies* 11: 3–6.

Blunt, A. and Wills, J. (2000) *Dissident Geographies: An Introduction to Radical Ideas and Practice*. Harlow: Prentice Hall.

Blunt, A., Gruffudd, P., May, J., Ogborn, M., and Pinder, D. (eds) (2003) *Cultural Geography in Practice*. London: Arnold.

Bookman, M. Z. (2002) *After Involuntary Migration: the Political Economy of Refugee Encampments*. Lanham, MD: Lexington Books.

Bowes, A. M., Dar, N. S., and Sim, D. F. (2002) 'Differentiation in housing careers: the case of Pakistanis in the UK'. *Housing Studies* 17: 381–99.

Bowlby, S., Gregory, S., and McKie, L. (1997) '"Doing home": patriarchy, caring and space'. *Women's Studies International Forum* 20(3): 343–50.

Brah, A. (1996) *Cartographies of Diaspora: Contesting Identities*. London: Routledge.

Brenton, M. (1998) *We're in Charge: CoHousing Communities of Older People in the Netherlands: Lessons for Britain?* Bristol: The Policy Press and The Housing Corporation.

Briganti, C. and Mezei, K. (2004) 'House haunting: the domestic novel of the inter-war years'. *Home Cultures* 1: 147–68.

Brinks, E. and Talley, L. (1996) 'Unfamiliar ties: lesbian constructions of home and family in Jeanette Winterson's *Oranges Are Not the Only Fruit* and Jewelle Gomez's *The Gilda Stories*'. In Wiley, C. and Barnes, F. R. (eds), *Homemaking: Women Writers and the Politics and Poetics of Home*. New York: Garland Publishing Inc., pp. 145–74.

British Parliamentary Papers (1887) *Report of the Royal Commission of the Colonial and Indian Exhibition, London, 1886*. London.

Brook, I. (2003) 'Making here like there: place attachment, displacement and the urge to garden'. *Ethics, Place and Environment* 6: 227–34.

Brown, M. (2000) *Closet Space: Geographies of Metaphor from the Body to the Globe*. London: Routledge.

Brown, M. (2003) 'Hospice and the spatial paradoxes of terminal care', *Environment and Planning A*, 35: 833–51.

Brown, M. and Colton, T. (2001) 'Dying epistemologies: an analysis of home death and its critique'. *Environment and Planning A* 33: 799–821.

Bryden, I. (2004) '"There is no outer without inner space": constructing the *haveli* as home'. *Cultural Geographies* 11: 26–41.

Bryden, I. and Floyd, J. (eds) (1999) *Domestic Space: Reading the Nineteenth-Century Interior*. Manchester: Manchester University Press.

Brydon, C. (1978) *The Lucknow Siege Diary of Mrs C. M. Brydon*, ed. and pub. C. De L. W. Fforde.

Bryson, L. and Winter, I. (1999) *Social Change, Suburban Lives: An Australian Newtown, 1960s to 1990s*. St Leonards: Allen and Unwin.

Buchli, V. and Lucas, G. (2001) *Archaeologies of the Contemporary Past*. London: Routledge.

Buck, N. *et al.* (eds) (1994) *Changing Households: the British Household Panel Survey 1990–1992*. Colchester: ESRC Research Centre on Micro-Social Change, University of Essex.

Buettner, E. (2004) *Empire Families: Britons in Late Imperial India*. Oxford: Oxford University Press.

Bunkše, E. (2004) *Geography and the Art of Life*. Baltimore, MD: The Johns Hopkins University Press.

Burn, G. (2004) 'Outdoors indoors'. *The Guardian*, 19 May.

Burton, A. (2003) *Dwelling in the Archive: Women Writing House, Home and History in Late Colonial India*. Oxford: Oxford University Press.

Buzar, S., Ogden, P. E., and Hall, R. (2005) 'Households matter: the quiet demography of urban transformation'. *Progress in Human Geography* 29: 413–36.

Buzzelli, M. (2004) 'Exploring regional firm-size structure in Canadian housebuilding: Ontario, 1991 and 1996', *Urban Geography* 25(3): 241–63.

Cairns, S. (ed.) (2004) *Drifting: Architecture and Migrancy*. London: Routledge.

Campo, J. E. (1991) *The Other Side of Paradise: Explorations into the Religious Meanings of Domestic Space in Islam*. Columbia, SC: University of South Carolina Press.

Carter, S. (2005) 'The geopolitics of diaspora'. *Area* 37: 54–63.

Carton, A. (2004) 'A passionate occupant of the transnational transit lounge'. In Kwan, S. and Speirs, K. (eds) *Mixing it Up: Multiracial Subjects*. Austin: University of Texas Press, pp. 73–90.

Case, A. (1858) *Day by Day at Lucknow: A Journal of the Siege of Lucknow*. London: Richard Bentley.

Castles, S. and Miller, M. J. (2003) *The Age of Migration: International Population Movements in the Modern World*. New York: Palgrave Macmillan.

Castree, N. (2005) *Nature*. London: Routledge.

Chambers, D. (1997) 'A stake in the country: women's experiences of suburban development'. In Silverstone, R. (ed.), *Visions of Suburbia*. London: Routledge, pp. 86–107.

Chambers, D. (2003) 'Family as place: family photograph albums and the domestication of public and private space'. In Schwartz, J. M. and Ryan, J. R. (eds), *Picturing Place: Photography and the Geographical Imagination*. London: IB Tauris, pp. 96–114.

Chambers, I. (1990) *Border Dialogues*. London: Routledge.

Chambers, I. (1993) *Migrancy, Culture, Identity*. London: Routledge.

Chant, S. (1997) *Women-Headed Households: Diversity and Dynamics in the Developing World*. Basingstoke: Macmillan.

Chapman, T. (2004) *Gender and Domestic Life: Changing Practices in Families and Households*. Basingstoke and New York: Palgrave Macmillan.

Chapman, T. and Hockey, J. (eds) (1999) *Ideal Homes? Social Change and Domestic Life*. London and New York: Routledge.

Chatterjee, P. (1993) *The Nation and its Fragments: Colonial and Postcolonial Histories*. New Delhi: Oxford University Press.

Chaudhuri, N. (1992) 'Shawls, curry and rice in Victorian Britain'. In Chaudhuri, N. and Strobel, M. (eds), *Western Women and Imperialism: Complicity and Resistance*. Bloomington: Indiana University Press, pp. 231–46.

Cheang, S. (2001) 'The Dogs of Fo: gender, identity and collecting'. In Shelton, A. (ed.), *Collectors: Expressions of Self and Other*. London: Horniman; Universidade de Coimbra.

Checkoway, B. (1980) 'Large builders, Federal Housing Programs, and Postwar Suburbanization'. *International Journal of Urban and Regional Research* 4(1): 21–45.

Cieraad, I. (1999a) 'Introduction: anthropology at home'. In Cieraad, I. (ed.) *At Home: an Anthropology of Domestic Space*. Syracuse: Syracuse University Press, pp. 1–12.

Cieraad, I. (1999b) 'Dutch windows: female virtue and female vice'. In Cieraad, I. (ed.), *At Home: an Anthropology of Domestic Space*. Syracuse: Syracuse University Press.

Clarke, A. J. (1997) 'Tupperware: suburbia, sociality and mass consumption'. In Silverstone, R. (ed.), *Visions of Suburbia*. London: Routledge, pp. 132–60.

Clarke, A. J. (2001a) *Tupperware: the Promise of Plastic in 1950s America*. Washington, DC: Smithsonian Books.

Clarke, A. J. (2001b) 'The aesthetics of social aspiration'. In Miller, D. (ed.), *Home Possessions: Material Culture behind Closed Doors*. Oxford: Berg, pp. 23–45.

Clifford, J. (1997) *Routes: Travel and Translation in the Late Twentieth Century*. Cambridge, MA: Harvard University Press.

Cloke, P., Philo, C. and Sadler, D. (1991) *Approaching Human Geography: An Introduction to Contemporary Theoretical Debates*. London: Chapman.

Collins, P. H. (1991) *Black Feminist Thought: Knowledge, Consciousness, and the Politics of Empowerment*. New York and London: Routledge.

Comaroff, J. and Comaroff, J. (1992) 'Home-made hegemony: modernity, domesticity, and colonialism in South Africa'. In Tranberg Hansen, K. (ed.), *African Encounters with Domesticity*. New Brunswick, NJ: Rutgers University Press, pp. 37–74.

Constable, N. (2003) 'Filipina workers in Hong Kong homes: household rules and relations', in Ehrenreich, B. and Hochschild, A. R. (eds), *Global Woman: Nannies, Maids, and Sex Workers in the New Economy*. New York: Metropolitan Books, pp. 115–41.

Cooper Marcus, C. (1995) *House as a Mirror of Self: Exploring the Deeper Meaning of Home*. Berkeley, CA: Conari Press.

Corbridge, S. (1999) '"The militarization of all Hindudom?" The Bharatiya Janata Party, the bomb, and the political spaces of Hindu nationalism'. *Economy and Society* 28: 222–55.

Costello, L. (2005) 'From prisons to penthouses: the changing images of high-rise living in Melbourne'. *Housing Studies* 20(1): 49–62.

Costello, L. and Hodge, S. (1999) 'Queer/clear/here: destabilising sexualities and spaces', in E. Stratford (ed.), *Australian Cultural Geographies*. Melbourne: Oxford University Press, pp. 131–52.

Cowen, D. (2004) 'From the American lebensraum to the American living room: class, sexuality, and the scaled production of "domestic" intimacy'. *Environment and Planning D: Society and Space* 22: 755–71.

Cox, R. and Narula, R. (2003) 'Playing happy families: rules and relationships in au pair employing households in London, England', *Gender, Place and Culture* 10(4): 333–44.

Crouch, D. (2004) 'Writing of Australian dwelling: animate houses and anxious ground'. *Journal of Australian Studies* 80: 43–52.

CSI (Colonization Society of India) (1935) 'McCluskiegunge: on the Ranchi Pateau'. Brochure held in the private archive of Alfred de Rozario, McCluskieganj.

Csikszentmihalyi, M. and Rochberg-Halton, E. (1981) *The Meaning of Things: Domestic Symbols and the Self*. Cambridge: Cambridge University Press.

Curtis, J. C. (1998) 'Race, realism, and the documentation of the rural home during America's Great Depression'. In Thompson, E. McD. (ed.), *The American Home: Material Culture, Domestic Space, and Family Life*. Winterthur, DE: Henry Francis du Pont Winterthur Museum, pp. 273–99.

Daniels, I. M. (2001) 'The "untidy" Japanese house'. In Miller, D. (ed.), *Home Possessions: Material Culture behind Closed Doors*. Oxford: Berg, pp. 201–29.

David, D. (1999) 'Imperial chintz: domesticity and empire'. *Victorian Literature and Culture* 27: 569–77.

Davidoff, L. and Hall, C. (2002) *Family Fortunes: Men and Women of the English Middle Class, 1780–1850*, London and New York: Routledge [first published in 1987 by Hutchinson Education].

Davin, A. (1978) 'Imperialism and motherhood'. *History Workshop Journal* 5: 9–65.

Deacon, R., Gormley, A., and Wilding, A. (2002) 'Mass-production, distribution and destination'. In Painter, C. (ed.), *Contemporary Art and the Home*. Oxford: Berg, pp. 181–94.

Depres, C. (1991) 'The meaning of home: literature review, directions for future research', *Journal of Architectural and Planning Research* 8(2): 96–115.

DeVault, M. (1991) *Feeding the Family: The Social Organization of Caring as Gendered Work*. Chicago: University of Chicago Press.

Diken, B. (2004) 'From refugee camps to gated communities: biopolitics and the end of the city'. *Citizenship Studies* 8: 83–106.

Diver, M. (1909) *The Englishwoman in India*. Edinburgh: Blackwood.

Dohmen, R. (2004) 'The home in the world: women, threshold designs and performative relations in contemporary Tamil Nadu, South India'. *Cultural Geographies* 11: 7–25.

Domosh, M. and Seager, J. (2001) *Putting Women in Place: Feminist Geographers Make Sense of the World*. New York: Guilford.

Dorai, M. K. (2002) 'The meaning of homeland for the Palestinian diaspora: revival and transformation'. In Al-Ali, N. and Koser, K. (eds), *New Approaches to Migration? Transnational Communities and the Transformation of Home*. London: Routledge, pp. 87–95.

Dovey, K. (1985) 'Home and homelessness', in Altman, I. and Werner, C. M. (eds), *Home Environments*. New York and London: Plenum Press, pp. 33–64.

Dovey, K. (1992) 'Model houses and housing ideology in Australia', *Housing Studies* 7(3): 177–88.

Dovey, K. (1999) *Framing Places: Mediating Power in Built Form*. London and New York: Routledge.

Dowling, R. (1998a) 'Suburban stories, gendered lives: thinking through difference'. In Fincher, R. and Jacobs, J. M. (eds), *Cities of Difference*. New York: Guilford.

Dowling, R. (1998b) 'Neotraditionalism in the suburban landscape: cultural geographies of exclusion in Canada', *Urban Geography* 19(2): 105–22.

Dowling, R. (2000) 'Power, ethics and subjectivity in qualitative research'. In Hay, I. (ed.), *Qualitative Methods in Human Geography*. Melbourne: Oxford University Press, pp. 23–36.

Dowling, R. and Mee, K. (2000) 'Tales of the city: western Sydney at the end of the millennium', in J. Connell (ed.), *Sydney: The Emergence of a World City*. Melbourne: Oxford University Press, pp. 273–91.

Duncan, J. S. (2002) 'Embodying colonialism? Domination and resistance in nineteenth century Ceylonese coffee plantations'. *Journal of Historical Geography* 28: 317–38.

Duncan, J. S. and Duncan, N. G. (2004) *Landscapes of Privilege: The Politics of the Aesthetic in an American Suburb*. New York and London: Routledge.

Duncan, J. S. and Lambert, D. (2003) 'Landscapes of home'. In Duncan, J. S., Johnson, N. C., and Schein, R. H. (eds), *A Companion to Cultural Geography*. Oxford: Blackwell, pp. 382–403.

Dwyer, C. (2002) '"Where are you from?" Young British Muslim women and the making of "home"', in Blunt, A. and McEwan, C. (eds), *Postcolonial Geographies*. London: Continuum, pp. 184–99.

Dwyer, D. and Bruce, J. (eds) (1988) *A Home Divided: Women and Income in the Third World*. Stanford, CA: Stanford University Press.

Dyck, I. (1990) 'Space, time and renegotiating motherhood: an exploration of the domestic workplace'. *Environment and Planning D: Society and Space* 8: 459–83.

Dyck, I., Kontos, P., Angus, J., and McKeever, P. (2005) 'The home as a site for long term care: meanings and management of bodies and spaces'. *Health and Place* 11(1), pp. 173–85.

Easthope, H. (2004) 'A place called home', *Housing, Theory and Society* 21(3): 128–38.

Eastmond, M. and Öjendal, J. (1999) 'Revisiting a "repatriation success": the case of Cambodia'. In Black, R. and Koser, K. (eds), *The End of the Refugee Cycle? Repatriation and Reconstruction*. Oxford: Berghahn Books, pp. 38–55.

Eidse, F. and Sichel, N. (2004) 'Introduction', in Eidse, F. and Sichel, N. (eds), *Unrooted Childhoods: Memoirs of Growing Up Global*. London: Nicholas Brearley Publishing, pp. 1–6.

Ellis, P. and Khan, Z. (2002) 'The Kashmiri diaspora: influences in Kashmir'. In Al-Ali, N. and Koser, K. (eds), *New Approaches to Migration? Transnational Communities and the Transformation of Home*. London: Routledge, pp. 169–85.

Elwood, S. (2000) 'Lesbian living spaces: multiple meanings of home'. *Journal of Lesbian Studies* 4: 11–28.

Eng, D. (1997) 'Out here and over there: queerness and diaspora in Asian American studies'. *Social Text* 15: 31–52.

Ephraums, E. (ed.) (2000) *The Big Issue Book of Home*. London: The Big Issue and Hodder and Stoughton.

Evans, N. (2005) 'Size matters'. *International Journal of Cultural Studies* 8(2): 131–49.

Fainstein, S. (1995) *City Builders: Property, Politics and Planning in London and New York*. Cambridge, MA: Blackwell.

Faley, J. (1990) *Up Oor Close: Memories of Domestic Life in Glasgow Tenements, 1910–1945*. Wendlebury: White Cockade.

Ferguson Ellis, K. (1989) *The Contested Castle: Gothic Novels and the Subversion of Domestic Ideology*. Urbana: University of Illinois Press.

Ferris Motz, M. and Browne, P. (eds) (1988) *Making the American Home: Middle Class Women and Domestic Material Culture, 1840–1940*. Bowling Green, OH: Bowling Green State University Popular Press.

Fincher, R. (2004) 'Gender and life course in the narratives of Melbourne's high-rise housing developers'. *Australian Geographical Studies* 42: 325–38.

Fincher, R. and Costello, L. (2003) 'Housing ethnicity: multicultural negotiation and housing the transnational student', in Yeoh, B. S. A., Charney, M. W., and Kiong, T. C. (eds), *Approaching Transnationalisms: Studies on Transnational Socieites, Multicultural Contacts, and Imaginings of Home*. Boston: Kluwer, pp. 161–86.

Flanagan, M. (2003) 'SIMple and personal: domestic space and the Sims'. *MelbourneDAC* 2003 (available online: http://hypertext.rmit.edu.au/dac/papers/Flanagan.pdf).

Floyd, J. (2002a) *Writing the Pioneer Woman*. Columbia: University of Missouri Press.

Floyd, J. (2002b) 'Domestication, domesticity and the work of butchery: positioning the writing of colonial housework'. *Women's History Review* 11: 395–415.

Floyd, J. (2004) 'Coming out of the kitchen: texts, contexts and debates'. *Cultural Geographies* 11: 61–72.

Forrest, R. and Lee, J. (2004) 'Cohort effects, differential accumulation and Hong Kong's volatile housing market', *Urban Studies* 41(11): 2181–96.

Forrest, R. and Murie, A. (1992) 'Change on a rural council estate: an analysis of dwelling histories', *Journal of Rural Studies* 8(1): 53–65.

Fortier, A.-M. (2000) *Migrant Belongings: Memory, Space, Identity*. Oxford: Berg.

Fortier, A.-M. (2001) '"Coming home": queer migrations and multiple evocations of home'. *European Journal of Cultural Studies* 4: 405–24.

Fortier, A.-M. (2003) 'Making home: queer migrations and motions of attachment'. In Ahmed, S., Castañeda, C., Fortier, A.-M., and Sheller, M. (eds), *Uprootings/Regroundings: Questions of Home and Migration*. Oxford: Berg, pp. 115–35.

Fouron, G. E. (2003) 'Haitian immigrants in the United States: the imagining of where "home" is in their transnational social fields'. In Yeoh, B. S. A., Charney, M. W., and Kiong, T. C. (eds), *Approaching Transnationalisms: Studies on Transnational Societies, Multicultural Contacts, and Imaginings of Home*. Boston: Kluwer Academic Publishers, pp. 205–50.

Frank, A. (1967) *The Diary of a Young Girl*. New York: Doubleday.

Freeman, J. (2004) *The Making of the Modern Kitchen: a Cultural History*. Oxford: Berg.

Friedan, B. (1963) *The Feminine Mystique*. Norton: New York.

Fromm, D. (1991) *Collaborative Communities: Cohousing, Central Living, and Other New Forms of Housing with Shared Facilities*. New York: Van Nostrand Reinhold.

Galsworthy, J. (1924) *The White Monkey*. London: Heinemann.

Ganguly, K. (1992) 'Migrant identities: personal memory and the construction of selfhood'. *Cultural Studies* 6: 27–50.

Garner, J. (2005) 'Editor's letter'. *The Big Issue*, South Africa, 90.

Gelder, K. and Jacobs, J. M. (1998) *Uncanny Australia: Sacredness and Identity in a Postcolonial Nation*. Carlton: Melbourne University Press.

George, R. M. (1996) *The Politics of Home: Postcolonial Relocations and Twentieth-Century Fiction*. Cambridge: Cambridge University Press.

George, R. M. (ed.) (1999) *Burning Down the House: Recycling Domesticity*. Boulder, CO: Westview Press.

Germon, M. (1957) *Journal of the Siege of Lucknow: An Episode of the Indian Mutiny*, ed. M. Edwardes. London: Constable.

Gibbs, A. and King, D. (2003) 'The electronic ball and chain? The operation and impact of home detention with electronic monitoring in New Zealand'. *The Australian and New Zealand Journal of Criminology* 36: 1–17.

Gilbert, M. R. (1994) 'The politics of location: doing feminist research at "home"'. *Professional Geographer* 46: 90–6.

Giles, J. (2004) *The Parlour and the Suburb: Domestic Identities, Class, Femininity and Modernity*. Oxford: Berg.

Gilman, C. P. (2002) [1903] *The Home, its Work and Influence*. Walnut Creek, CA: AltaMira Press.

Gilroy, P. (1993) *The Black Atlantic: Modernity and Double Consciousness*. London: Verso.

Glover, W. J. (2004) '"A feeling of absence from Old England": the colonial bungalow'. *Home Cultures* 1: 61–82.

Gluck, S. B. and Patai, D. (eds) (1991) *Women's Words: the Feminist Practice of Oral History*. New York: Routledge.

Goh, R. B. H. (2001) 'Ideologies of "upgrading" in Singapore public housing: post-modern style, globalisation and class construction in the built environment'. *Urban Studies* 38(9):1589–604.

Goldsmith, B. (1999) 'All quiet on the Western front? Suburban reverberations in recent Australian cinema'. *Australian Studies* 14: 115–32.

Gooder, H. and Jacobs, J. M. (2002) 'Belonging and non-belonging: the apology in a reconciling nation', in Blunt, A. and McEwan, C. (eds) *Postcolonial Geographies*. London: Continuum, pp. 200–13.

Gorman-Murray, A. (2006) 'Homeboys: uses of home by gay Australian men'. *Social and Cultural Geographies* 7: 53–69.

Gowans, G. (2001) 'Gender, imperialism and domesticity: British women repatriated from India, 1940–1947'. *Gender, Place and Culture* 8: 255–69.

Gowans, G. (2002) 'A passage from India: geographies and experiences of repatriation, 1858–1939'. *Social and Cultural Georgraphy* 3: 403–23.

Gowans, G. (2003) 'Imperial geographies of home: memsahibs and miss-sahibs in India and Britain, 1915–1947'. *Cultural Geographies* 10: 424–41.

Gram-Hanssen, K. and Bech-Danielsen, C. (2004) 'House, home and identity from a consumption perspective'. *Housing, Theory and Society* 21: 17–26.

Gregson, N. and Lowe, M. (1994) *Servicing the Middle Classes*. London: Macmillan.

Greig, A. (1995) *The Stuff Dreams Are Made Of: Housing Provision in Australia 1945–1960*. Melbourne University Press: Carlton.

Grewal, I. (1996) *Home and Harem: Nation, Gender, Empire, and the Cultures of Travel*. Durham, NC: Duke University Press.

Gullestad, M. (1984) *Kitchen-table Society: A Case Study of the Family Life and Friendships of Young Working-class Mothers in Urban Norway*. Oslo: Columbia University Press.

Gurney, C. (1997) '" ... Half of me was satisfied": making sense of home through episodic ethnographies'. *Women's Studies International Forum* 20: 373–86.

Gurney, C. M. (1999a) 'Pride and Prejudice: discourses of normalisation in public and private accounts of home ownership'. *Housing Studies* 14(2): 163–83.

Gurney, C. M. (1999b) '"We've got friends who live in council houses": power and resistance in home ownership', in Hearn, J. and Roseneil, S. (eds), *Consuming Cultures: Power and Resistance*. London: Macmillan, pp. 42–68.

Haar, S. and Reed, C. (1996) 'Coming home: a postscript on postmodernism'. In Reed, C. (ed.), *Not at Home: the Suppression of Domesticity in Modern Art and Architecture*. London: Thames and Hudson, pp. 253–73.

Hage, G. (1996) 'The spatial imaginary of national practices: dwelling-domesticating/being-exterminating'. *Environment and Planning D: Society and Space* 14: 463–86.

Hage, G. (1997) 'At home in the entrails of the West: multiculturalism, ethnic food and migrant home-building'. In Grace, H., Hage, G., Johnson, L., Langsworth, J., and Symonds, M., *Home/World: Space, Community and Marginality in Sydney's West*. Sydney: Pluto.

Hall, C. (1992) *White, Male and Middle Class: Explorations in Feminism and History*. Cambridge: Polity Press.

Hall, S. (1987) 'Minimal selves'. In Appignanesi, L. (ed.), *Identity. The Real Me. Post-Modernism and the Question of Identity*. ICA Documents 6. London: Institute of Contemporary Arts.

Hall, S. (1996) 'When was the "postcolonial"? Thinking at the limit'. In Chambers, I. and Curti, L. (eds), *The Post-colonial Questions: Common Skies, Divided Horizons*. London: Routledge, pp. 242–60.

Halle, D. (1993) *Inside Culture: Art and Class in the American Home*. Chicago: University of Chicago Press.

Hammerton, A. J. and Thomson, A. (2005) *Ten Pound Poms: Australia's Invisible Migrants*. Manchester: Manchester University Press.

Hammond, L. C. (2004) *This Place will become Home: Refugee Repatriation to Ethiopia*. Ithaca, NY: Cornell University Press.

Hand, M. and Shove, E. (2004) 'Orchestrating concepts: kitchen dynamics and regime change in *Good Housekeeping* and *Ideal Home*, 1922–2002'. *Home Cultures* 1: 235–56.

Hanson, S. and Pratt, G. (1995) *Gender, Work and Space.* London and New York: Routledge.

Hanson, S. and Pratt, G. (2003) 'Learning about labour: combining qualitative and quantitative methods'. In Blunt, A. *et al.* (eds), *Cultural Geography in Practice*. London: Arnold, pp. 106–18.

Hapner, K. (2005) 'For victims, news about home can come from strangers online'. *The New York Times*, 5 September: 1.

Hardhill, I. (2004) 'Transnational living and moving experiences: intensified mobility and dual-career households'. *Population, Space and Place* 10: 375–89.

Hardy, D. (2000) *Utopian England: Community Experiments, 1900–1945*. London: E. and F.N. Spon.

Harper, M. (2005) 'Introduction'. In Harper, M. (ed.), *Emigrant Homecomings: The Return Movement of Emigrants, 1600–2000*. Manchester: Manchester University Press, pp. 1–14.

Harris, C. (2002) *Making Native Space: Colonialism, Resistance, and Reserves in British Columbia*. Vancouver: University of British Columbia Press.

Harris, K. (1858) *A Lady's Diary of the Siege of Lucknow*. London: John Murray.

Harris, R. (1996) *Unplanned Suburbs: Toronto's American Tragedy 1900 to 1950*. London and Baltimore, MD: Johns Hopkins University Press.

Harvey, D. (1978) 'Labor, capital and class struggle around the built environment in advanced capitalist societies', in K. Cox (ed.), *Urbanization and Conflict in Market Societies*. Chicago: Maaroufa, pp. 9–37.

Häusermann Fábos, A. (2002) 'Sudanese identity in diaspora and the meaning of home: the transformative role of Sudanese NGOs in Cairo'. In Al-Ali, N. and Koser, K. (eds), *New Approaches to Migration? Transnational Communities and the Transformation of Home.* London: Routledge, pp. 34–50.

Hay, I. (ed.) (2000) *Qualitative Research Methods in Human Geography.* Melbourne: Oxford University Press.

Hayden, D. (1996) [1981] *The Grand Domestic Revolution: A History of Feminist Designs for American Homes, Neighbourhoods, and Cities.* Cambridge MA: MIT Press.

Hayden, D. (2002) [1984] *Redesigning the American Dream: The Future of Housing, Work and Family Life.* New York: W. W. Norton.

Hayden, D. (2003) *Building Suburbia: Green Fields and Urban Growth, 1820–2000.* New York: Pantheon.

Hayden, D. with Wark, J. (2004) *A Field Guide to Sprawl.* New York: W. W. Norton.

Hays, C. L. (2005) 'Home sweet home confinement'. *The New York Times* 3 May: C1.

Heidegger, M. (1993) [1978] 'Building Dwelling Thinking', in D. F. Krell (ed.), *Basic Writings from* Being and Time *(1927) to* The Task of Thinking *(1964).* London: Routledge, pp. 347–63.

Herbert, S. (2000) 'For ethnography'. *Progress in Human Geography* 24: 550–68.

HIA (Housing Industry Association of Australia) (2004) *Housing 100.* Canberra: Housing Industry Association.

Hitchings, R. (2003) 'People, plants and performance: on actor network theory and the material pleasures of the private garden'. *Social and Cultural Geography* 4: 99–113.

Hitchings, R. (2004) 'At home with someone nonhuman'. *Home Cultures* 1: 169–86.

Hitchings, R. and Jones, V. (2004) 'Living with plants and the exploration of botanical encounter within human geographic research practice'. *Ethics, Place and Environment* 7: 3–18.

Ho, E. S. (2002) 'Multi-local residence, transnational networks: Chinese "Astronaut" families in New Zealand'. *Asian and Pacific Migration Journal* 11(1): 145–64.

Hochschild, A. (1989) *The Second Shift: Working Parents and the Revolution at Home.* New York: Viking.

Hochschild, A. R. (2003) 'Love and gold', in Ehrenreich, B. and Hochschild, A. R. (eds), *Global Woman: Nannies, Maids, and Sex Workers in the New Economy.* New York: Metropolitan Books, pp. 15–30.

Hockey, J. (1999) 'The ideal of home: domesticating the institutional space of old age and death'. In Chapman, T. and Hockey, J. (eds), *Ideal Homes? Social Change and Domestic Life.* London: Routledge, pp. 108–18.

Hodgkin, K. and Radstone, S. (eds) (2003a) *Contested Pasts: the Politics of Memory.* London: Routledge.

Hodgkin, K. and Radstone, S. (eds) (2003b) *Regimes of Memory.* London: Routledge.

Hoffman, E. (1989) *Lost in Translation: a Life in a New Language.* New York: Penguin.

Hoganson, K. (2002) 'Cosmopolitan domesticity: importing the American dream, 1865–1920', *American Historical Review*, February: 55–83.

Hoggart, K., Lees, L., and Davies, A. (2002) *Researching Human Geography.* London: Arnold.

Holloway, S. L. and Valentine, G. (2001) 'Children at home in the wired world: reshaping and rethinking the home in urban geography'. *Urban Geography* 22: 562–83.

Holloway, S. R. and Wyly, E. K. (2001) '"The color of money" expanded: geographically contingent mortgage lending in Atlanta'. *Journal of Housing Research* 12(1): 55–90.

hooks, b. (1991) *Yearning: Race, Gender, and Cultural Politics.* London: Turnaround.

Hope, C. (2003) 'Great white hope'. *The Guardian*, 6 May.

Houston, V. H. (1996) 'Home'. In Wiley, C. and Barnes, F. (eds), *Homemaking: Women Writers and the Politics and Poetics of Home*. New York: Garland Publishing, pp. 277–82.

Humble, N. (2001) *The Feminine Middlebrow Novel, 1920s–1950s: Class, Domesticity and Bohemianism*. Oxford: Oxford University Press.

Hyndman, J. (2000) *Managing Displacement: Refugees and the Politics of Humanitarianism*. Minneapolis: University of Minnesota Press.

Ifekwunigwe, J. (1999) *Scattered Belongings: Cultural Paradoxes of 'Race,' Nation and Gender*. London: Routledge.

Imrie, R. (2004a) 'Housing quality, disability and domesticity'. *Housing Studies* 19: 685–90.

Imrie, R. (2004b) 'Disability, embodiment and the meaning of the home'. *Housing Studies* 19: 745–64.

Imrie, R. (2005) *Accessible Housing: Quality, Disability and Design*. London: Routledge.

Inglis, J. (1892) *The Siege of Lucknow: A Diary*. London: James R. Osgood.

Irigaray, L. (1992) *Ethics of Sexual Difference*. Ithaca, NY: Cornell University Press.

Ironmonger, D. (1996) 'Counting outputs, capital inputs and caring labor: estimating gross household product', *Feminist Economies* 2(3): 37–64.

Jacobs, J. M. (1996) *Edge of Empire: Postcolonialism and the City*. London: Routledge.

Jacobs, J. M. (2001) 'Hybrid highrises', in Barrett, J. and Butler-Bowdon, C. (eds), *Debating the City*. Sydney: Historic Houses Trust of New South Wales, pp. 13–27.

Jacobs, J. M. (2004) 'Too many houses for a home: narrating the house in the Chinese diaspora'. In Cairns, S. (ed.), *Drifting: Architecture and Migrancy*. London: Routledge, pp. 164–83.

Jacobs, J. M. (2006) 'A geography of big things'. *Cultural Geographies* 13(1): 1–27.

Jacobs, K. (2001) 'Historical perspectives and methodologies: their relevance for housing studies?'. *Housing, Theory and Society* 18: 127–35.

Jacobs, K., Kemeny, J., and Manzi, T. (2003) 'Power, discursive space and institutional practices in the construction of housing problems'. *Housing Studies* 18(4): 429–46.

Johnson, L. (1993) 'Text-ured brick: speculations on the cultural production of domestic space'. *Australian Geographical Studies* 31: 201–13.

Johnson, L. (1994) 'Occupying the suburban frontier: accommodating difference on Melbourne's urban fringe'. In Blunt, A. and Rose, G. (eds), *Writing Women and Space: Colonial and Postcolonial Geographies*. New York: Guilford, pp. 141–68.

Johnson, L. (1996) '"As housewives we are worms": women, modernity and the home question'. *Cultural Studies* 10(3): 449–63.

Johnson, L. and Lloyd, J. (2004) *Sentenced to Everyday Life: Feminism and the Housewife*. Oxford: Berg.

Johnston, L. and Valentine, G. (1995) 'Wherever I lay my girlfriend, that's my home: the performance and surveillance of lesbian identities in domestic environments', in Bell, D. and Valentine, G. (eds), *Mapping Desire: Geographies of Sexualities*. London: Routledge, pp. 99–113.

Kaika, M. (2004) 'Interrogating the geographies of the familiar: domesticating nature and constructing the autonomy of the modern home'. *International Journal of Urban and Regional Research* 28: 265–86.

Kaplan, A. (2002) *The Anarchy of Empire in the Making of US Culture*. Cambridge, MA: Harvard University Press.

Kaplan, A. (2003) 'Homeland insecurities: reflections on language and space'. *Radical History Review* 85: 82–93.

Katz, C. (1993) 'Growing girls/closing circles: limits on the spaces of knowing in rural Sudan and United States cities', in Katz, C. and Monk, J. (eds), *Full Circles: Geographies of Women over the Life Course*. London: Routledge, pp. 88–106.

Katz, C. (1994) 'Playing the field: questions of fieldwork in geography'. *The Professional Geographer* 46: 67–72.

Katz, C. (2005) *Growing up Global: Economic Restructuring and Children's Everyday Lives.* Minneapolis: University of Minnesota Press.

Katz-Hyman, M. (1998) '"In the middle of this poverty some cups and a teapot": the furnishing of slave quarters at colonial Williamsburg'. In Thompson, E. McD. (ed.), *The American Home: Material Culture, Domestic Space, and Family Life.* Winterthur, DE: Henry Francis du Pont Winterthur Museum, pp. 197–216.

Kay, J. (1997) 'Sweet surrender, but what's the gender? Nature and the body in the writings of nineteenth-century Mormon women', in Jones, J. P., Nast, H., and Roberts, S. (eds), *Thresholds in Feminist Geography: Difference, Methodology, Representation.* Lanham, MD: Rowman and Littlefield, pp. 361–82.

Kaye, J. W. (1876) *A History of the Sepoy War in India, 1857–58.* London: W. H. Allen.

Kellett, P. and Moore, J. (2003) 'Routes to home: homelessness and home-making in contrasting societies'. *Habitat International* 27: 123–41.

Kelly, B. M. (1993) *Expanding the American Dream.* Albany: State University of New York Press.

Kendall, C. (1995) 'Appendix: link-up', in MacDonald, R., *Between Two Worlds: The Commonwealth Government and the Removal of Aboriginal Children of Part Descent in the Northern Territory.* Alice Springs: IAD Press, pp. 72–4.

Kennedy, D. (1996) *The Magic Mountains: Hill Stations and the British Raj.* Berkeley: University of California Press.

Kenny, J. (1995) 'Climate, race and imperial authority: the symbolic landscape of the British hill station in India'. *Annals of the Association of American Geographers* 85: 694–714.

Kenyon, L. (1999) 'A home from home: students' transitional experience of home', in Chapman, T. and Hockey, J. (eds), *Ideal Homes? Social Change and Domestic Life.* London: Routledge, pp. 84–95.

Kerber, L. K. (1988) 'Separate spheres, female worlds, woman's place: the rhetoric of women's history'. *The Journal of American History,* June: 9–39.

Kettl, D. F. (2003) 'Promoting state and local government performance for homeland security'. *The Century Foundation Homeland Security Project:* www.tcf.org.

Kibria, N. (2002) 'Of blood, belonging, and homeland trips: transnationalism and identity among second-generation Chinese and Korean Americans', in Levitt, P. and Waters, M. C. (eds), *The Changing Face of Home: the Transnational Lives of the Second Generation.* New York: Russell Sage Foundation, pp. 295–311.

King, A. (1984) *The Bungalow.* London: Routledge.

King, A. (1997) 'Excavating the multicultural suburb: hidden histories of the bungalow', in Silverstone, R. (ed.), *Visions of Suburbia.* London: Routledge, pp. 55–85.

King, A. D. and Kusno, A. (2000) 'On Be(ij)ing in the World: "Postmodernism," "Globalization," and the Making of Transnational Space in China', in A. Dirlik and X. Zhang (eds), *Postmodernism and China.* London and Durham, NC: Duke University Press: pp. 41–67.

King, P. (2004) 'The room to panic: an example of film criticism and housing research'. *Housing, Theory and Society* 21: 27–35.

Kneafsey, M. and Cox, R. (2002) 'Food, gender and Irishness: how Irish women in Coventry make home'. *Irish Geography* 35: 6–15.

Kneale, J. (2003)' Secondary worlds: reading novels as geographical research', in Blunt, A., Gruffudd, P., May, J., Ogborn, M., and Pinder, D. (eds), *Cultural Geography in Practice.* London: Arnold, pp. 39–51.

Kolodny, A. (1975) *The Lay of the Land: Metaphor as Experience and History in American Life and Letters.* Chapel Hill: University of North Carolina Press.

Kolodny, A. (1984) *The Land Before Her: Fantasy and Experience of the American Frontier, 1630–1860*. Chapel Hill: University of North Carolina Press.

Kranidis, R. S. (1999) *The Victorian Spinster and Colonial Emigration: Contested Subjects*. New York: St Martin's.

Kuczynski, A. (2002) 'When home is a castle and the big house, too'. *The New York Times*, 18 August.

Kuhn, A. (1995) *Family Secrets: Acts of Memory and Imagination*. London: Verso.

Landy, M. (2004) *Semi-Detached*. London: Tate Publishing.

Langland, E. (1995) *Nobody's Angels: Middle-class Women and Domestic Ideology in Victorian Culture*. Ithaca, NY: Cornell University Press.

Law, L. (2001) 'Home cooking: Filipino women and geographies of the senses in Hong Kong', *Ecumene* 8(3): 264–83.

Leach, R. (2002) 'What happened at home with art: tracing the experience of consumers'. In Painter, C. (ed.), *Contemporary Art and the Home*. Oxford: Berg, pp. 153–80.

Lee, R. (2005) 'Reconstructing "home" in apartheid Cape Town: African women and the process of settlement'. *Journal of Southern African Studies* 31(3): 611–30.

Legg, S. (2003) 'Gendered politics and nationalised homes: women and the anti-colonial struggle in Delhi, 1940–1947'. *Gender, Place and Culture* 10: 7–28.

Leisch, H. (2002) 'Gated communities in Indonesia', *Cities* 19(5): 341–50.

Leslie, D. (1993) 'Femininity, post-Fordism and the "new traditionalism"'. *Environment and Planning D: Society and Space* 11: 689–708.

Leslie, D. and Reimer, S. (2003) 'Gender, modern design and home consumption'. *Environment and Planning D: Society and Space* 21: 293–316.

Lester, A. (2001) *Imperial Networks: Creating Identities in Nineteenth-Century South Africa and Britain*. London: Routledge.

Ley, D. (1995) 'Between Europe and Asia: The case of the missing sequoias', *Ecumene* 2(2): 185–210.

Lim, S. (1997) *Among the White Moon Faces: an Asian-American Memoir of Homelands*. New York: Feminist Press at the CUNY.

Lingwood, J. (ed.) (1995) *Rachel Whiteread House*. London: Phaidon, in association with Artangel.

Listokin, D. *et al.* (2003) 'Known facts or reasonable assumptions? An examination of alternative sources of housing data'. *Journal of Housing Research* 13: 219–51.

Liu, M. (2004) 'Woes of a do-gooder: Beijing has effectively silenced SARS whistle-blower Jiang Yanyong using an old tool, house arrest'. *Newsweek International*, 18 October: 48.

Llewellyn, M. (2004a) '"Urban village" or "white house": envisioned spaces, experienced places, and everyday life at Kensal House, London in the 1930s'. *Environment and Planning D: Society and Space* 22: 229–249.

Llewellyn, M. (2004b) 'Designed by women and designing women: gender, planning and the geographies of the kitchen in Britain, 1917–1946'. *Cultural Geographies* 11: 42–60.

Lloyd, J. and Johnson, L. (2004) 'Dream stuff: the postwar home and the Australian housewife, 1940–60'. *Environment and Planning D: Society and Space* 22: 251–72.

Lowenthal, D. (1989). 'Nostalgia tells it like it wasn't', in Shaw, C. and Case, M. (eds), *The Imagined Past: History and Nostalgia*. Manchester: Manchester University Press, pp. 18–32.

Lower East Side Tenement Museum (2004) [1999] *A Tenement Story: the History of 97 Orchard Street and the Lower East Side Tenement Museum*. New York: Lower East Side Tenement Museum.

Luxton, M. (1997) 'The UN, women and household labour: measuring and valuing unpaid work', *Women's Studies International Forum* 20(3): 431–39.

Lyons, L. (1996) 'Feminist articulations of the nation: the "dirty" women of Armagh and the discourse of Mother Ireland'. *Genders* 24: 110–49.

McCamant, K. and Durrett, C. (1994) [1988] *Cohousing: a Contemporary Approach to Housing Ourselves*. Berkeley, CA: Ten Speed Press.

McClintock, A. (1993) 'Family feuds: gender, nationalism and the family'. *Feminist Review* 44: 61–80.

McClintock, A. (1995) *Imperial Leather: Race, Gender and Sexuality in the Colonial Contest.* New York: Routledge.

McCloud, A. B. (1996) '"This is a Muslim home": signs of difference in the African-American row house', in Metcalf, B. D. (ed.), *Making Muslim Space in North America and Europe.* Berkeley: University of California Press, pp. 65–73.

MacDonald, R. (1995) *Between Two Worlds: The Commonwealth Government and the Removal of Aboriginal Children of Part Descent in the Northern Territory.* Alice Springs: IAD Press.

McDowell, L. (1999) 'Scales, spaces and gendered differences: a comment on gender cultures', *Geoforum* 30: 231–3.

McDowell, L. (2000) 'The trouble with men? Young people, gender transformations and the crisis of masculinity'. *International Journal of Urban and Regional Research* 24: 201–9.

McDowell, L. and Sharp, J. P. (eds) (1999) *A Feminist Glossary of Human Geography.* London: Arnold.

McGrath, M. (2002) *Silvertown: An East End Family Memoir.* London: Fourth Estate.

Mack, J. (2004) 'Inhabiting the imaginary: factory women at home on Batam Island, Indonesia'. *Singapore Journal of Tropical Geography* 25: 156–79.

Mackenzie, S. and Rose, D. (1983) 'Industrial change, the domestic economy and home life', in Anderson, J., Duncan, S., and Hudson, R. (eds), *Redundant Spaces in Cities and Regions.* London: Academic Press, pp. 175–6.

McNamee, S. (1998) 'Youth, gender and video games: power and control in the home', in Skelton, T. and Valentine, G. (eds), *Cool Places: Geographies of Youth Culture.* London: Routledge, pp. 195–206.

McNeill, D. (2005) 'In search of the global architect: the case of Norman Foster (and Partners)'. *International Journal of Urban and Regional Research* 29: 501–15.

Madigan, R. and Munro, M. (1991) 'Gender, house and home: social meanings and domestic architecture in Britain'. *Journal of Architecture and Planning Research* 8: 116–32.

Madigan, R. and Munro, M. (1996) 'House beautiful: style and consumption in the home'. *Sociology* 30: 41–57.

Madigan, R. and Munro, M. (1999a) 'The more we are together: domestic space, gender and privacy', in Chapman, T. and Hockey, J. (eds), *Ideal Homes? Social Change and Domestic Life.* London and New York: Routledge, pp. 61–72.

Madigan, R. and Munro, M. (1999b) 'Negotiating space in the family home', in I. Cieraad (ed.), *At Home: An Anthropology of Domestic Space.* New York: Syracuse University Press, pp. 107–17.

Magnusson, L. and Özüekren, A. S. (2002) 'The housing careers of Turkish households in middle-sized Swedish municipalities'. *Housing Studies* 17: 465–86.

Mahtani, M. (2002) 'What's in a name? Exploring the employment of "mixed race" as an identification'. *Ethnicities* 2: 469–90.

Mallett, S. (2004) 'Understanding home: a critical review of the literature', *The Sociological Review* 52(1): 62–89.

Manning, E. (2003) *Ephemeral Territories: Representing Nation, Home, and Identity in Canada.* Minneapolis: University of Minnesota Press.

Manzo, L. (2003) 'Beyond house and haven: toward a revisioning of emotional relationships with places'. *Journal of Environmental Psychology* 23: 47–61.

Marcus, C. C. (1999) *House as a Mirror of Self: Exploring the Deeper Meaning of Home*, Berkeley, CA: Conari Press.

Marston, S. A. (2000) 'The social construction of scale', *Progress in Human Geography* 24(2): 219–42.

Marston, S. A. (2004) 'A long way from home: domesticating the social production of scale'. In Sheppard, E. and McMaster, R. (eds), *Scale and Geographic Inquiry: Nature, Society and Method*. Oxford: Blackwell, pp. 170–91.

Martin, B. and Mohanty, C. T. (1986) 'Feminist politics: what's home got to do with it?', in de Lauretis, T. (ed.), *Feminist Studies/Critical Studies*. Basingstoke: Macmillan.

Mass-Observation (1943) *An Enquiry into People's Homes*. London: John Murray.

Massey, D. (1991) 'A global sense of place'. *Marxism Today*, June: 24–9.

Massey, D. (1992) 'A place called home'. *New Formations* 17: 3–15.

Massey, D. (1995a) 'Masculinity, dualisms and high technology', *Transactions, Institute of British Geographers* 20(4): 487–99.

Massey, D. (1995b) 'Space-time and the politics of location'. In Lingwood, J. (ed.), *Rachel Whiteread House*. London: Phaidon Press, in association with Artangel, pp. 34–49.

Massey, D. (2001) 'Living in Wythenshawe', in Borden, I., Kerr, J., and Rendell, J. with Pivaro, A. (eds), *The Unknown City: Contesting Architecture and Social Space*. Boston: MIT Press, pp. 458–75.

Massey, D. (2005) *For Space*. London: Sage.

May, J. (2000a) 'Housing histories and homeless careers: a biographical approach'. *Housing Studies* 15: 613–38.

May, J. (2000b) 'Of nomads and vagrants: single homelessness and narratives of home as place'. *Environment and Planning D: Society and Space* 18: 737–59.

Medina, J. (2005) 'Far from home, the survivors of two hurricanes become neighbours, mentors and friends'. *The New York Times*, 30 September: 20.

Mee, K. J. (1993) 'Roll up, roll up! It's the greatest show in town'. In *IAG Conference Proceedings*, Monash Publications in Geography, pp. 207–21.

Meluish, C. (2005) 'Michael Landy's *Semi-Detached*'. *Home Cultures* 2: 117–22.

Messerschmidt, D. A. (ed.) (1981) *Anthropology at Home in North America: Methods and Issues in the Study of One's Own Society*. Cambridge: Cambridge University Press.

Meth, P. (2003) 'Rethinking the "domus" in domestic violence: homelessness, space and domestic violence in South Africa'. *Geoforum* 34: 317–28.

Mezei, K. and Briganti, C. (2002) 'Reading the house: a literary perspective'. *Signs* 27: 837–46.

Mifflin, E. and Wilton, R. (2005) 'no place like home: rooming houses in contemporary urban context'. *Environment and Planning A* 37: 403–21.

Miller, D. (1988) 'Appropriating the state on the council estate', *MAN* 23: 353–72.

Miller, D. (1998) 'Why some things matter', in Miller, D. (ed.), *Material Cultures: Why Some Things Matter*. London: University College London Press, pp. 3–21.

Miller, D. (ed.) (2001) *Home Possessions: Material Culture behind Closed Doors*. Oxford: Berg.

Miller, D. (2002) 'Accommodating', in Painter, C. (ed.), *Contemporary Art and the Home*. Oxford and New York: Berg, pp. 115–30.

Mitchell, K. (2004) 'Conflicting landscapes of dwelling and democracy in Canada', in Cairns, S. (ed.), *Drifting: Architecture and Migrancy*. London: Routledge, pp. 142–64.

Moreton-Robinson, A. (2000) *Talkin' up to the White Woman: Indigenous Women and Feminism*. St Lucia: Queensland University Press.

Moreton-Robinson, A. (2003) 'I still call Australia home: indigenous belonging and place in a white postcolonizing society', in Ahmed, S., Castañeda, C., Fortier, A.-M., and Sheller, M. (eds), *Uprootings/Regroundings: Questions of Home and Migration.* Oxford: Berg, pp. 23–40.

Morgan, G., Rocha, C., and Poynting, S. (2005) 'Grafting cultures: longing and belonging in immigrants' gardens and backyards in Fairfield'. *Journal of Intercultural Studies* 26(1–2): 93–105.

Morgan, S. (1990) *My Place.* Sydney: Pan Books.

Morley, D. (1986) *Family Television: Cultural Power and Domestic Leisure.* Comedia: London.

Morley, D. (2000) *Home Territories: Media, Mobility and Identity.* London: Routledge.

Mortimer, L. (2000) '*The Castle*, the garbage bin and the high-voltage tower: home truths in the suburban grotesque'. *Metro* 123: 8–12.

Moss, P. (1997) 'Negotiating spaces in home environments: older women living with arthritis'. *Social Science and Medicine* 45: 23–33.

Moss, P. (ed.) (2000) *Placing Autobiography in Geography.* Syracuse: Syracuse University Press.

Murdie, R. (2002) 'The housing careers of Polish and Somali newcomers in Toronto's rental market'. *Housing Studies* 17: 423–43.

Murray, M. (1995) 'Correction at Cabrini-Green: a sociospatial exercise of power', *Environment and Planning D: Society and Space* 13: 311–27.

Muzzio, D. and Halper, T. (2002) 'Pleasantville? The suburb and its representation in American movies'. *Urban Affairs Review* 37: 543–74.

Myers, J. C. (2001) 'Performing the voyage out: Victorian female emigration and the class dynamics of displacement'. *Victorian Literature and Culture* 29: 129–46.

Myerson, J. (2004) *Home: the Story of Everyone who Ever Lived in Our House.* London: Flamingo.

Nagar, R., Lawson, V., McDowell, L., and Hanson, S. (2002) 'Locating globalisation: femininst (re) readings of the subjects and spaces of globalisation'. *Economic Geography* 78(3): 257–84.

Nash, C. (2000) 'Performativity in practice: some recent work in cultural geography'. *Progress in Human Geography* 24: 653–64.

Nesbitt, J. (2004) 'Everything must go'. In Landy, M., Nesbitt, J. and Slyce, J. (eds), *Michael Landy: Semi-Detached.* London: Tate Publishing, pp. 12–49.

Neuwirth, R. (2004) *Shadow Cities: A Billion Squatters, A New Urban World.* London and New York: Routledge.

Newman, K. and Wyly, E. K. (2004) 'Geographies of mortgage market segmentation: the case of Essex County, New Jersey'. *Housing Studies* 19(1): 53–83.

Nicolaides, B. (1999) '"Where the working man is welcomed": working-class suburbs in Los Angeles, 1900–1940'. *Pacific Historical Review* 68: 517–59.

Noble, G. (2002) 'Comfortable and relaxed: furnishing the home and nation'. *Continuum: Journal of Media and Cultural Studies* 16: 53–66.

Oberhauser, A. M. (1997) 'The home as "field": households and homework in rural Appalachia', in Jones, J. P. et al. (eds), *Thresholds in Feminist Geography: Difference, Methodology, Representation.* Lanham, MD: Rowman and Littlefield, pp. 165–82.

ODPM (2004) *Table 201 Housebuilding: Permanent Dwellings Started and Completed, by Tenure, United Kingdom.* Available at www.odpm.gov.uk.

Ogden, P. E. and Hall, R. (2004) 'The second demographic transition, new household forms and the urban population of France in the 1990s'. *Transactions* 29: 88–105.

Ogden, P. E. and Schnoebelen, F. (2005) 'The rise of the small household: demographic change and household structure in Paris'. *Population, Space and Place* 11: 251–68.

Oldman, C. and Beresford, B. (2000) 'Home, sick home: using the housing experiences of disabled children to suggest a new theoretical framework', *Housing Studies* 15(3): 429–42.

Olds, K. (1995) 'Globalization and the production of new urban spaces: Pacific Rim megaprojects in the late 20th century'. *Environment and Planning A* 27(11): 1713–43.

Olds, K. (2001) *Globalization and Urban Change: Capital, Culture, and Pacific-Rim Mega-Projects*. Oxford: Oxford University Press.

Oliver, P. (1987) *Dwellings: The House across the World*. Oxford: Phaidon Press.

Olumide, J. (2002) *Raiding the Gene Pool: the Social Construction of Mixed Race*. London: Pluto Press.

Oncu, A. (1997) 'The myth of the ideal home travels across cultural borders to Istanbul', in Oncu, A. and Weyland, P. (eds), *Space, Culture and Power: New Identities in Globalizing Cities*. London and Atlantic Highlands, NJ: Zed Books, pp. 56–72.

Østergaard-Nielson, E. (2002) 'Working for a solution through Europe: Kurdish political lobbying in Germany', in Al-Ali, N. and Koser, K. (eds), *New Approaches to Migration? Transnational Communities and the Transformation of Home*. London: Routledge, pp. 186–201.

Øverland, O. (2005) 'Visions of home: exiles and immigrants', in Rose, P. I. (ed.), *The Dispossessed: An Anatomy of Exile*. Amherst: University of Massachusetts Press, pp. 7–26.

Özüekren, A. S. and Van Kempen, R. (2002) 'Housing careers of minority ethnic groups: experiences, explanations and prospects'. *Housing Studies* 17: 365–79.

Pain, R. (1997) 'Social geographies of women's fear of crime'. *Transactions* 22: 231–44.

Painter, C. (1999) *At Home with Art*. London: Hayward Gallery Publishing.

Painter, C. (2002a) 'Introduction', in Painter, C. (ed.), *Contemporary Art and the Home*. Oxford: Berg, pp. 1–6.

Painter, C. (2002b) *Contemporary Art and the Home*. Oxford: Berg.

Painter, C. (2002c) 'The *At Home with Art* project: a summary', in Painter, C. (ed.), *Contemporary Art and the Home*. Oxford: Berg, pp. 7–9.

Papastergiadis, N. (1996) *Dialogues in the Diaspora: Essays and Conversations on Cultural Identity*. London: Rivers Oram Press.

Parr, J. (1999) *Domestic Goods: the Material, the Moral, and the Economic in the Postwar Years*. Toronto: University of Toronto Press.

Pateman, C. (1989) *The Disorder of Women: Democracy, Feminism, and Political Theory*, Stanford, CA: Stanford University Press.

Pearlman, M. (1996) *A Place Called Home: Twenty Writing Women Remember*. New York: St Martin's Press.

Pearson, R. (2004) 'Organising home-based workers in the global economy: an action-research approach'. *Development in Practice* 14: 136–48.

Peel, M. (2003) *The Lowest Rung: Voices of Australian Poverty*. Melbourne: Cambridge University Press.

Percival, J. (2002) 'Domestic spaces: uses and meanings in the daily lives of older people', *Ageing and Society* 22: 729–49.

Peres Da Costa, S. (1999) 'On homesickness: narratives of longing and loss in the writings of Jamaica Kincaid'. *Postcolonial Studies* 2: 75–89.

Perks, R. and Thomson, A. (eds) (1998) *The Oral History Reader*. London: Routledge.

Perry, D. (1983) *Backtalk: Women Writers Speak Out: Interviews by Donna Perry*. Piscataway, NJ: Rutgers University Press.

Peters, J. D. (1999) 'Exile, nomadism, and diaspora: the stakes of mobility in the western canon'. In Naficy, H. (ed.), *Home, Exile, Homeland: Film, Media, and the Politics of Place*. New York: Routledge, pp. 17–41.

Phillips, R. (1997) *Mapping Men and Empire: a Geography of Adventure*. London: Routledge.

Pickles, K. (2002a) 'Pink cheeked and surplus: single British women's inter-war migration to New Zealand', in Fraser, L. and Pickles, K. (eds), *Shifting Centres: Women and Migration in New Zealand History*. Dunedin: University of Otago Press, pp. 63–80.

Pickles, K. (2002b) *Female Imperialism and National Identity: Imperial Order Daughters of the Empire*. Manchester: Manchester University Press.

Pilkington, H. and Flynn, M. (1999) 'From "refugee" to "repatriate": Russian repatriation discourse in the making', in Black, R. and Koser, K. (eds), *The End of the Refugee Cycle? Repatriation and Reconstruction*. Oxford: Berghahn Books, pp. 171–97.

Pilkington, M. [Nugi Garimara] (1996) *Follow the Rabbit-Proof Fence*. St Lucia, Queensland: University of Queensland Press.

Pinder, D. (2005) *Visions of the City*. Edinburgh: Edinburgh University Press.

Pink, S. (2004) *Home Truths: Gender, Domestic Objects and Everyday Life*. New York: Berg.

Platt, K. (1923) *The Home and Health in India and the Tropical Colonies*. London: Bailliere, Tindall and Co.

Pollock, D. and Van Reken, R. (1999) *Third Culture Kids: the Experience of Growing Up Among Worlds*. London: Intercultural Press.

Pollock, G. (1988) *Vision and Difference: Femininity, Feminism and the Histories of Art*. London: Routledge.

Porteous, J. D. and Smith, S. E. (2001) *Domicide: the Global Destruction of Home*. Montreal: McGill-Queen's University Press.

Posonby, M. (2003) 'Ideals, reality and meaning: homemaking in England in the first half of the nineteenth century'. *Journal of Design History* 16: 201–14.

Postle, M., Daniels, S. and Alfrey, N. (eds) (2004) *Art of the Garden*. London: Tate Publishing.

Povrzanovic Frykman, M. (2002) 'Homeland lost and gained: Croatian diaspora and refugees in Sweden', in Al-Ali, N. and Koser, K. (eds), *New Approaches to Migration? Transnational Communities and the Transformation of Home*. London: Routledge, pp. 118–37.

Power, E. (2005) 'Human-nature relations in suburban gardens'. *Australian Geographer* 36: 39–53.

Pratt, G. (1981) 'The house as an expression of social worlds', in Duncan, J. S. (ed.), *Housing and Identity: Cross-cultural Perspectives*. London: Croom Helm, pp. 135–75.

Pratt, G. (1987) 'Class, home, and politics'. *Canadian Review of Sociology and Anthropology* 24: 39–55.

Pratt, G. (1997) 'Stereotypes and ambivalence: the construction of domestic workers in Vancouver, British Columbia', *Gender, Place and Culture* 4(2): 159–77.

Pratt, G. (1998) 'Geographic metaphors in feminist theory', in Aiken, S. H. *et al.* (eds), *Making Worlds: Gender, Metaphor, Materiality*. Tucson: University of Arizona Press, pp. 13–30.

Pratt, G. (2003) 'Valuing childcare: troubles in suburbia', *Antipode* 35(3): 581–602.

Pratt, G. (2004) *Working Feminism*. Edinburgh: Edinburgh University Press.

Pratt, M. B. (1984) 'Identity: skin blood heart', in Burkin, E., Pratt, M. B., and Smith, B. (eds), *Yours in Struggle: Three Feminist Perspectives on Anti-Semitism and Racism*. Ithaca, NY: Firebrand Books, pp. 9–64.

Procida, M. (2002) *Married to the Empire: Gender, Politics and Imperialism in India, 1883–1947*. Manchester: Manchester University Press.

Radcliffe, S. A. (1996) 'Gendered nations: nostalgia, development and territory in Ecuador'. *Gender, Place and Culture* 3: 5–22.

Rakoff, R. M. (1977) 'Ideology in everyday life: the meaning of the house', *Politics and Society* 7: 85–104.

Ramirez-Machado, J. M. (2004) *Domestic Work, Conditions of Work and Employment: A Legal Perspective*, Conditions of Work and Employment Series No. 7. Geneva: International Labor Office.

Randall, G. (1988) *No Way Home: Homeless Young People in Central London*. London: Centrepoint.

Rapoport, A. (1995) 'A critical look at the concept "home"', in Benjamin, D. N., Stea, D., and Saile, D. (eds), *The Home: Words, Interpretations, Meanings and Environments*. Aldershot: Avebury, pp. 25–52.

Rapport, N. and Dawson, A. (eds) (1998) *Migrants of Identity: Perceptions of Home in a World of Movement*. Oxford: Berg.

Ratcliffe, P. (1998) '"Race", housing and social exclusion', *Housing Studies* 13(6): 807–18.

Ravetz, A. (1995) *The Place of the Home: English Domestic Environments, 1914–2000*. London: Spon.

Read, P. (1996) *Returning to Nothing: the Meaning of Lost Places*. Cambridge: Cambridge University Press.

Read, P. (2000) *Belonging: Australians, Place and Aboriginal Ownership*. Cambridge: Cambridge University Press.

Reed, C. (ed.) (1996) *Not at Home: the Suppression of Domesticity in Modern Art and Architecture*. London: Thames and Hudson.

Reed, C. (2002) 'Domestic disturbances: challenging the anti-domestic modern', in Painter, C. (ed.), *Contemporary Art and the Home*. Oxford: Berg, pp. 35–54.

Reiger, K. M. (1985) *The Disenchantment of the Home: Modernising the Australian Family 1880–1940*. Melbourne: Oxford University Press.

Reimer, S. and Leslie, D. (2004) 'Identity, consumption and the home'. *Home Cultures* 1: 187–208.

Relph, E. (1976) *Place and Placelessness*. London: Pion.

Rhode, D. and Dewan, S. (2005) 'More deaths confirmed in homes for the aged'. *The New York Times*, 15 September: 21.

Richards, T. (1990) *The Commodity Culture of Victorian Britain: Advertising and Spectacle, 1851–1914*. London: Verso.

Ritchie, W. K. (1997) *Miss Toward of the Tenement House*. Edinburgh: The National Trust for Scotland.

Robinson, C. (2002) '"I think home is more than a building": young home(less) people on the cusp of home, self and something else'. *Urban Policy and Research* 20(1): 27–38.

Rodriguez, I. (1994) *Home, Garden, Nation: Space, Gender and Ethnicity in Postcolonial Latin American Literatures by Women*. Durham, NC: Duke University Press.

Romines, A. (1992) *The Home Plot: Women, Writing and Domestic Ritual*. Amherst: University of Massachusetts Press.

Rose, G. (1993) *Feminism and Geography: The Limits of Geographical Knowledge*. Cambridge: Polity.

Rose, G. (2001) *Visual Methodologies: An Introduction to the Interpretation of Visual Materials*. London: Sage.

Rose, G. (2003) 'Family photographs and domestic spacings: a case study'. *Transactions* 28: 5–18.

Rose, G. (2004) '"Everyone's cuddled up and it just looks really nice": an emotional geography of some mums and their family photos'. *Social and Cultural Geography* 5: 549–64.

Rubenstein, R. (2001) *Home Matters: Longing and Belonging, Nostalgia and Mourning in Women's Fiction*. New York: Palgrave.

Ryan, D. S. (1997a) *The Ideal Home Through the Twentieth Century*. London: Hazar Publishing.

Ryan, D. S. (1997b) 'The empire at home: the Daily Mail Ideal Home Exhibition and the imperial suburb'. Imperial Cities Project Working Paper 6, Department of Geography, Royal Holloway, University of London.

Ryan, N. (2003) *Homeland: Into a World of Hate*. Edinburgh: Mainstream.

Rybczynski , W. (1988) *Home: A Short History of an Idea*. London: Heinemann.

Said, E. (1978) *Orientalism*. New York: Penguin.

Said, E. (1993) *Culture and Imperialism*. New York: Vintage.

Salih, R. (2002) 'Shifting meanings of "home": consumption and identity in Moroccan women's transnational practices between Italy and Morocco'. In Al-Ali, N. and Koser, K. (eds), *New Approaches to Migration? Transnational Communities and the Transformation of Home*. London: Routledge, pp. 51–67.

Saunders, P. and Williams, P. (1988) 'The constitution of the home: towards a research agenda', *Housing Studies* 3(2): 81–93.

Schaffer, K. (1988) *Women and the Bush: Forces of Desire in the Australian Cultural Tradition*. Cambridge: Cambridge University Press.

Schneir, M. (ed.) (1996) *The Vintage Book of Historical Feminism*. London: Vintage.

Schwartz-Cowan, R. (1989) *More Work for Mother: The Ironies of Household Technology from the Open Hearth to the Microwave*. London: Free Association Books.

Sciorra, J. (1996) 'Return to the future: Puerto Rican vernacular architecture in New York City', in King, A. D. (ed.), *Re-Presenting the City: Ethnicity, Capital and Culture in the 21st-Century Metropolis*. London: Macmillan, pp. 60–90.

Scott, S. (2004) 'Transnational exchanges amongst skilled British migrants in Paris', *Population, Space and Place* 10: 391–410.

Seed, P. (1995) *Ceremonies of Possession in Europe's Conquest of the New World, 1492–1640*. Cambridge: Cambridge University Press.

Sharp, J. (2000) 'Towards a critical analysis of fictive geographies'. *Area* 32: 327–34.

Sharpe, J. (1993) *Allegories of Empire: the Figure of Woman in the Colonial Text*. Minneapolis: University of Minnesota Press.

Shaw, C. and Case, M. (eds) (1989) *The Imagined Past: History and Nostalgia*. Manchester: Manchester University Press.

Shiach, M. (2005) 'Modernism, the city and the "domestic interior"'. *Home Cultures* 2: 251–68.

Shohat, E. (1999) 'By the bitstream of Babylon: cyberfrontiers and diasporic vistas', in Naficy, H. (ed.), *Home, Exile, Homeland: Film, Media, and the Politics of Place*. New York: Routledge, pp. 213–32.

Shove, E. (2003) *Comfort, Cleanliness and Convenience: The Social Organization of Normality*. Oxford and New York: Berg.

Sichel, N. (2004) 'Going home', in Eidse, F. and Sichel, N. (eds), *Unrooted Childhoods: Memoirs of Growing Up Global*. London: Nicholas Brearley Publishing, pp. 185–98.

Silverstone, R. (1994) *Television and Everyday Life*. London and New York: Routledge.

Silvey, R. (2000a) 'Diasporic subjects: gender and mobility in Sulawesi'. *Women's Studies International Forum* 23: 501–15.

Silvey, R. (2000b) 'Stigmatized spaces: moral geographies under crisis in south Sulawesi, Indonesia'. *Gender, Place and Culture* 7: 143–61.

Silvey, R. (2003) 'Gender and mobility: critical ethnographies of migration in Indonesia', in Blunt, A., Gruffudd, P., May, J., Ogborn, M., and Pinder, D. (eds), *Cultural Geography in Practice*. London: Arnold, pp. 91–102.

Sinfield, A. (2000) 'Diaspora and hybridity. Queer identity and the ethnicity model'. In Mirzoeff, N. (ed.), *Diaspora and Visual Culture: Representing Africans and Jews*. London: Routledge.

Skelton, T. and Valentine, G. (1998) *Cool Places: Geographies of Youth Cultures*. London and New York: Routledge.

Smart, A. (2002) 'Agents of eviction: the squatter control and clearance division of Hong Kong's housing department'. *Singapore Journal of Tropical Geography* 23(3): 333–47.

Smith, B. H. (1988) *Contingencies of Value: Alternative Perspectives for Critical Theory*. Cambridge, MA: Harvard University Press.

Smith, J. A. (2003) 'Beyond dominance and affection: living with rabbits in post-humanist households'. *Society and Animals* 11(2): 181–97.

Somerville, P. (1992) 'Homelessness and the meaning of home: rooflessness or rootlessness?' *International Journal of Urban and Regional Research* 16(4): 529–39.

Southerton, D., Shove, E., Warde, A., and Deem, R. (2001) 'The social worlds of caravaning: objects, scripts and practices'. *Sociological Research Online* 6(2) (available online: www.socresonline.org.uk/6/2/southerton.html).

Spark, C. (1999) 'Home on "The Block": rethinking Aboriginal emplacement', in F. Murphy and E. Warner (eds), '*New Talents 21C Writing Australia*', *Journal of Australian Studies* 63: 56–63. Available online at www.api-network.com/pdf/jas63_spark.pdf.

Spark, C. (2003) 'Documenting Redfern: representing home and Aboriginality on the Block'. *Continuum: Journal of Media and Cultural Studies* 17: 33–50.

Sparke, P. (1995) '*As Long as it's Pink': The Sexual Politics of Taste*. London: Pandora.

Spence, J. and Holland, P. (eds) (1991) *Family Snaps: the Meanings of Domestic Photography*. London: Virago.

Spigel, L. (2001) 'Media homes then and now'. *International Journal of Cultural Studies* 4: 385–411.

Staeheli, L. *et al.* (2002) 'Immigration, the internet, and the spaces of politics'. *Political Geography* 21: 989–1012.

Statistics Canada (2004) *Table: Housing Stats, by province*. Available at www.statcan.ca (accessed 23 March 2006).

Steel, F. A. and Gardiner, G. (1907) *The Complete Indian Housekeeper and Cook*. 5th edition. London: Heinemann.

Stewart, S. (1993) *On Longing: Narratives of the Miniature, the Gigantic, the Souvenir, the Collection*. Durham, NC: Duke University Press.

Stoler, A. L. (1995) *Race and the Education of Desire: Foucault's History of Sexuality and the Colonial Order of Things*. Durham, NC: Duke University Press.

Stoler, A. L. (2002) *Carnal Knowledge and Imperial Power: Race and the Intimate in Colonial Rule*. Berkeley: University of California Press.

Sudjic, D. and Beyerle, T. (1999) *Home: the Twentieth-Century House*. London: Laurence King.

Tachibana, S., Daniels, S., and Watkins, C. (2004) 'Japanese gardens in Edwardian Britain: landscape and transculturation'. *Journal of Historical Geography* 30: 364–94.

Tasca, L. (2004) 'The "average housewife" in post-World War II Italy'. Translated by Stuart Hilwig. *Journal of Women's History* 16: 92–115.

Thapar-Björkert, S. (1997) 'The domestic sphere as a political site: a study of women in the Indian nationalist movement'. *Women's Studies International Forum* 20: 493–504.

Thapar-Björkert, S. and Ryan, L. (2002) 'Mother India/Mother Ireland: comparative gendered dialogues of colonialism and nationalism in the early 20th century'. *Women's Studies International Forum* 25: 301–13.

Thompson, B. and Tyagi, S. (eds) (1996) *Names we Call Home: Autobiography on Racial Identity*. New York: Routledge.

Thompson, E. McD. (ed.) (1998) *The American Home: Material Culture, Domestic Space, and Family Life*. Winterthur, DE: Henry Francis du Pont Winterthur Museum.

Thompson, P. (2000) *The Voice of the Past: Oral History.* 3rd edition. Oxford: Oxford University Press.

Thompson, S. (1994) 'Suburbs of opportunity: the power of home for migrant women', in Gibson, K. and Watson, S. (eds), *Metropolis Now: Planning and the Urban in Contemporary Australia.* Leichhardt, NSW: Pluto Press, pp. 33–45.

Thomson, A. (2005) '"My wayward heart": homesickness, longing and the return of British post-war immigrants from Australia', in Harper, M. (ed.), *Emigrant Homecomings: The Return Movement of Emigrants, 1600–2000.* Manchester: Manchester University Press, pp. 105–30.

Tickly, L., Caballero, C., Haynes, J. and Hill, J. (2004) *Understanding the Educational Needs of Mixed Heritage Pupils.* Research Brief RB549. London: Department for Education and Skills.

Tillman, F. (2000) 'A prologue to middle age', in Ephraums, E. (ed.), *The Big Issue Book of Home.* London: The Big Issue and Hodder and Stoughton, p. 81.

Tizard, B. and Phoenix, A. (2002) *Black, White, or Mixed Race? Race and Racism in the Lives of Young People of Mixed Parentage.* New York: Routledge.

Tolia-Kelly, D. (2004a) 'Materializing post-colonial geographies: examining the textural landscapes of migration in the South Asian home'. *Geoforum* 35: 675–88.

Tolia-Kelly, D. (2004b) 'Locating processes of identification: studying the precipitates of re-memory through artefacts in the British Asian home'. *Transactions* 29: 314–29.

Tolia-Kelly, D. (2006) 'Mobility/stability: British Asian cultures of "landscape and Englishness". *Environment & Planning A 38: 341–58.*

Tong, R. (1989) *Feminist thought: a comprehensive introduction.* Boulder, CO: Westview Press.

Tosh, J. (1995) 'Imperial masculinity and the flight from domesticity, 1880–1914', in Foley, T. P., Pilkington, L., Ryder, S., and Tilley, E. (eds), *Gender and Colonialism.* Galway: Galway University Press, pp. 72–85.

Tosh, J. (1999) *A Man's Place: Masculinity and the Middle-Class Home in Victorian England.* London: Routledge.

Tristram, P. (1989) *Living Space in Fact and Fiction.* London: Routledge.

Tuan, Y.-F. (1971) 'Geography, phenomenology and the study of human nature', *Canadian Geographer* 15: 181–92.

Turkington, R. (1999) 'British "corporation suburbia": the changing fortunes of Norris Green, Liverpool'. In Harris, R. and Larkham, P. J. (eds), *Changing Suburbs: Foundation, Form and Function.* London: E&FN Spon, pp. 56–75

Tuson, P. (1998) 'Mutiny narratives and the imperial feminine: European women's accounts of the rebellion in India in 1857'. *Women's Studies International Forum* 21: 291–303.

UNCHS/Habitat (2000) *Strategies to Combat Homelessness.* United Nations Centre for Human Settlements, Nairobi.

US Census Bureau (2004) *American Fact Finder: New Housing Stats.* Available at http://factfinder.census.gov (accessed 23 March 2006).

Valentine, G. (1993) '(Hetero)sexing space: lesbian perceptions and experiences of everyday spaces'. *Environment and Planning D: Society and Space* 11(4): 395–413.

Valentine, G. (1997) '"My son's a bit dizzy, my wife's a bit soft": gender, children and cultures of parenting'. *Gender, Place and Culture* 4: 37–62.

Valentine, G. (2001a) *Social Geographies: Society and Space.* Longman: Harlow.

Valentine, G. (2001b) *Stranger Danger.* London: Continuum.

Van Chaudenberg, A. and Heynen, H. (2004) 'The rational kitchen in the interwar period in Belgium: discourses and realities'. *Home Cultures* 1: 23–50.

van der Horst, H. (2004) 'Living in a reception centre: the search for home in an institutional setting.' *Housing, Theory and Society* 21: 36–46.

Van Hear, N. (2002) 'Sustaining societies under strain: remittances as a form of transnational exchange in Sri Lanka and Ghana'. In Al-Ali, N. and Koser, K. (eds), *New Approaches to Migration? Transnational Communities and the Transformation of Home.* London: Routledge, pp. 202–23.

Varley, A. (1996) 'Women heading households: some more equal than others?'. *World Development* 24: 506–20.

Varley, A. (2002) 'Gender, families and households', in Desai, V. and Potter, R. B. (eds), *The Companion to Development Studies.* London: Arnold, pp. 329–34.

Varley, A. and Blasco, M. (2001) 'Exiled to the home: masculinity and ageing in urban Mexico'. *European Journal of Development Research* 12(2):115–38.

Veness, A. R. (1993) 'Neither homed nor homeless: contested definitions and the personal worlds of the poor'. *Political Geography* 12(4): 319–40.

Veness, A. R. (1994) 'Design shelters as models and makers of home: new responses to homelessness in urban America'. *Urban Geography* 15(2): 150–67.

Vickery, A. (1993) 'Golden age to separate spheres? A review of the categories and chronology of English women's history'. *The Historical Review* 36: 383–414.

Vidler, A. (1994) *The Architectural Uncanny: Essays in the Modern Unhomely.* Cambridge, MA: MIT Press.

Waetjen, T. (1999) 'The "home" in homeland: gender, national space, and Inkatha's politics of ethnicity'. *Ethnic and Racial Studies* 22: 653–78.

Waitt, G. and Markwell, K. (2006) *Gay Tourism: Culture and Context.* New York: Haworth Press.

Wakeley, M. (2003) *Dream Home.* Allen and Unwin: Crows Nest.

Walker, A. (1984) *In Search of Our Mothers' Gardens.* London: The Women's Press.

Walker, L. (2002) 'Home making: an architectural perspective'. *Signs* 27: 823–35.

Wallis, A. D. (1997) *Wheel Estate: The Rise and Decline of Mobile Homes.* London and Baltimore, MD: Johns Hopkins University Press.

Walsh, J. (2004) *Domesticity in Colonial India: What Women Learned When Men Gave Them Advice.* Lanham, MD: Rowman and Littlefield Publishers.

Walter, B. (1995) 'Irishness, gender and place'. *Environment and Planning D: Society and Space* 13: 35–50.

Walter, B. (2001) *Outsiders Inside: Whiteness, Place and Irish Women.* London: Routledge.

Walters, W. (2004) 'Secure borders, safe haven, domopolitics'. *Citizenship Studies* 8: 237–60.

Warrington, M. (2001) '"I must get out": the geographies of domestic violence'. *Transactions* 26: 365–82.

Waters, J. (2002) 'Flexible families? "Astronaut" households and the experiences of lone mothers in Vancouver, British Columbia'. *Social & Cultural Geography* 3(2): 117–34.

Watkins, H. (1997) 'Kettle's Yard, Cambridge: Art House and Way of Life'. Unpublished MA dissertation, Royal Holloway, University of London.

Watkins, H. (2006) 'Beauty queen, bulletin board and browser: rescripting the refrigerator'. *Gender, Place and Culture* 13: 143–52.

Web Japan (2004) *Japan Web Statistics: Housing Stats.* Available at http://web-jpn.org/stat/index.html (accessed 23 March 2006).

Webster, W. (1998) *Imagining Home: Gender, 'Race' and National Identity, 1945–1964.* London: University College London Press.

Western, J. (1992) *A Passage to England: Barbadian Londoners Speak of Home.* London: University College London Press.

Weston, K. (1995) 'Get thee to a big city: sexual imaginary and the great gay migration'. *GLQ* 2: 253–77.

Westwood, S. and Phizacklea, A. (2000) *Trans-nationalism and the Politics of Belonging*. London: Routledge.

Whatmore, S. (2002) *Hybrid Geographies: Natures, Cultures and Spaces*. London: Sage.

White, P. and Hurdley, L. (2003) 'International migration and the housing market: Japanese corporate movers in London', *Urban Studies* 40(4): 687–706.

Wickham, C. J. (1999) *Constructing Heimat in Post-War Germany: Longing and Belonging*. Lewiston, NY: The Edwin Mellen Press.

Wiese, A. (1999) 'The other suburbanites: African American suburbanization in the North before 1950'. *The Journal of American History* 85(4): 1495–524.

Wiley, C. and Barnes, F. R. (eds) (1996) *Homemaking: Women Writers and the Politics and Poetics of Home*. New York: Garland Publishing.

Wilson, A. C. (1904) *Hints for the First Years of Residence in India*. Oxford: Clarendon Press.

Wilson, F. (2004) 'A house of several stories'. *The Guardian*, 15 May.

Wolch, J. and Dear, M. (1993) *Malign Neglect: Homelessness in an American City*. San Francisco: Jossey-Bass Publishers.

Wolf, D. (2002) 'There's no place like "home": emotional transnationalism and the struggles of second-generation Filipinos', in Levitt, P. and Waters, M. C. (eds), *The Changing Face of Home: the Transnational Lives of the Second Generation*. New York: Russell Sage Foundation, pp. 255–94.

Wood, D. and Beck, R. J., with Wood, I., Wood, R., and Wood, C. (1994) *Home Rules*. Baltimore, MD: Johns Hopkins University Press.

Woolf, V. (1945) [1928] *A Room of One's Own*. London: Penguin.

Wu, F. (2004) 'Transplanting cityscapes: the use of imagined globalisation in housing commodification in Beijing'. *Area* 36(3): 227–34.

Wu, F. (2005) 'Rediscovering the "gate" under market transition: from work-unit compounds to commodity housing enclaves'. *Housing Studies* 20(2): 235–354.

Wu, F. and Webber, K. (2004) 'The rise of "foreign gated communities" in Beijing: between economic globalization and local institutions'. *Cities* 21(3): 203–13.

Wyman, M. (2005) 'Emigrants returning: the evolution of a tradition', in Harper, M. (ed.), *Emigrant Homecomings: the Return Movement of Emigrants, 1600–2000*. Manchester: Manchester University Press, pp. 16–31.

Yates, J. (2002) 'Housing implications of social, spatial and structural change', *Housing Studies* 17(4): 581–618.

Yeoh, B. S. A. and Huang, S. (2000) '"Home" and "away": foreign domestic workers and negotiations of diasporic identity in Singapore'. *Women's Studies International Forum* 23(4): 413–29.

Yeoh, B. S. A., Charney, M. W. and Kiong, T. C. (eds) (2003) *Approaching Transnationalism: Studies on Transnational Societies, Multicultural Contacts, and Imaginings of Home*. Boston: Kluwer Academic Publishers.

Young, I. M. (1997) 'House and home: feminist variations on a theme', in *Intersecting Voices: Dilemmas of Gender, Political Philosophy, and Policy*. Princeton, NJ: Princeton University Press, pp. 134–64.

Zandy, J. (ed.) (1990) *Calling Home: Working Class Women's Writings*. New Brunswick, NJ: Rutgers University Press.

Zernike, K. and Wilgoren, J. (2005) 'In search of a place to sleep, and news of home'. *The New York Times* 31 August: 1.

Zlotnick, S. (1995) 'Domesticating imperialism: curry and cookbooks in Victorian England'. *Frontiers: a Journal of Women's Studies* 16: 51–68.

LIST OF WEBSITES

Listed in order of occurence in the main text.

CHAPTER 2

www.oralhistory.org.uk (Oral History Society)

www.hidden-histories.org/esch_pages (Eastside Community Heritage, London)

www.massobs.org.uk (Mass-Observation Archives, University of Sussex)

http://migration.ucc.ie/oralarchive (Irish Centre for Migration Studies, University College Cork)

www.umaine.edu/wic/both/FOHP (Maine Feminist Oral History Project)

www.bobbybakersdailylife.com (website of performance artist, Bobby Baker)

www.lgihome.co.uk (Home, Camberwell, London)

www.history.org (Colonial Williamsburg, Virginia, USA)

www.homethestory.com (Website for J. Myerson (2004) *Home: the Story of Everyone who ever Lived in our House.* London: Flamingo)

www.tenement.org (Lower East Side Tenement Museum, New York City)

http://hearth.library.cornell.edu/h/hearth (Home Economics Archive: Research, Tradition and History, Cornell University, USA)

www.artangel.co.uk (Artangel, London, UK)

www.artistsineastlondon.org (Artists in East London)

www.geog.ucl.ac.uk/~dtkelly/publications_media.htm (Melanie Carvalho's landscapes of home)

www.rabbit.org (House Rabbit Society)

CHAPTER 3

www.habitat.org (Habitat for Humanity)

www.womensaid.org.uk (Women's Aid, UK)

www.ifrc.org (International Federation of Red Cross and Red Crescent Societies)

www.uic.edu/~pbhales/Levittown.html (Levittown: Documents of an Ideal American Suburb. Website compiled by Peter Bacon Hales, Art History Department, University of Illinois, Chicago)

www.civictrust.org.uk (The Civic Trust, UK)

www.peabody.org.uk (The Peabody Trust, UK)

www.bigissue.com (The *Big Issue*, UK)

www.bigissue.org.za (The *Big Issue*, South Africa)

CHAPTER 4

www.dhs.gov (Department of Homeland Security, USA)

http://homelandsecurity.osu.edu/NACHS (National Academic Consortium for Homeland Security, USA)

http://homelandsecurityinstitute.org (Homeland Security Institute, USA)

www.esri.com (ESRI: GIS and Mapping Software)

www.dfat.gov.au/facts/separated_children.html (Department of Foreign Affairs and Trade, Australian Government)

www.austlii.edu.au/au/special/rsjproject/
rsjlibrary/hreoc/stolen (Bringing Them
Home Report, 1997, on the website of the
Reconciliation and Social Justice Library,
Australia)

www.nla.gov.au (National Library of
Australia)

www.cohre.org (Centre on Housing Rights
and Evictions)

www.intermix.org.uk (Intermix, UK)

CHAPTER 5

www.unhcr.ch (Office of the United Nations
High Commissioner for Refugees)

www.hact.org.uk (Housing Associations'
Charitable Trust, UK)

http://england.shelter.org.uk (Shelter, UK)

www.refugeehousing.org.uk (Refugee
Housing, UK)

www.anglo-indians.com (Website on the
Anglo-Indian community, compiled and
managed by Bert Payne, USA)

www.alphalink.com.au/~agilbert (Website
on the Anglo-Indian community,
compiled and managed by Adrian Gilbert,
Australia, which publishes two electronic
journals)

www.cfmw.org/files/migrantwomen.htm
(Commission for Filipino Migrant Workers:
'A House is not a Home' programme)

www.hrw.org/reports/2001/usdom/
usadom0501–01.htm (Human Rights
Watch: 'Hidden in the Home')

http://ourworld.compuserve.com/
homepages/kalayaan/home.htm
(KALAYAAN – Justice for Overseas
Domestic Workers)

www.apwld.org/lm.htm (Labour and
Migration Task Force, Asia Pacific Forum
on Women, Law and Development:
focus on Domestic Work, 2003–5)

www.ips-dc.org/campaign (The Institute
for Policy Studies, USA: Break the Chain
Campaign)

www.udwa.org (United Domestic Workers
of America)

www.caaav.org (Domestic Workers United)

CHAPTER 6

http://news.bbc.co.uk (BBC News)

http://usatoday.com (USA Today)

www.redcross.org (Red Cross, USA)

http://news.independent.co.uk (The Independent, UK)

www.cohousing.org (The Cohousing Association of the United States)

www.cohousingresources.com (Co-
Housing Resources LLC, USA)

www.cohousing.co.uk/owch.htm (Older
Women's Cohousing, London, UK)

Index